1st International Congress on Coastal Research (ICCR 2020)

E3S Web of Conferences Volume 298 (2021)

Al Hoceima, Morocco
6-9 July 2021

Editors:

Said El Moussaoui
Hajar El Talibi
Hinde Cherkaoui Dekkaki
Issam Etebaai

ISBN: 978-1-7138-3487-8

Printed from e-media with permission by:

Curran Associates, Inc.
57 Morehouse Lane
Red Hook, NY 12571

Some format issues inherent in the e-media version may also appear in this print version.

For additional information, please contact EDP Sciences – Web of Conferences
at the address below.

EDP Sciences – Web of Conferences
17, Avenue du Hoggar
Parc d'Activité de Courtabœuf
BP 112
F-91944 Les Ulis Cedex A
France

Phone: +33 (0) 1 69 18 75 75
Fax: +33 (0) 1 69 28 84 91

contact-edps@webofconferences.org

Additional copies of this publication are available from:

Curran Associates, Inc.
57 Morehouse Lane
Red Hook, NY 12571 USA
Phone: 845-758-0400
Fax: 845-758-2633
Email: curran@proceedings.com
Web: www.proceedings.com

TABLE OF CONTENTS

COASTAL HAZARDS AND RISK ASSESSMENT

INTEGRATED COASTAL ZONE MANAGEMENT

OCEANOGRAPHY, COASTAL ECOLOGY, AND RESOURCES

GEOMORPHOLOGY AND COASTAL PROCESSES

COASTAL HYDROLOGY AND HYDROGEOLOGY

edp sciences

E3S Web of Conferences

Statement of Peer review

In submitting conference proceedings to E3S Web of Conferences, I certify to the Publisher that I adhere to the **Policy on Publishing Integrity** of the journal in order to safeguard good scientific practice in publishing.

1. All articles have been subjected to peer review administered by the proceedings editors.
2. Reviews have been conducted by expert referees, who have been requested to provide unbiased and constructive comments aimed, whenever possible, at improving the work.
3. Proceedings editors have taken all reasonable steps to ensure the quality of the materials they publish and their decision to accept or reject a paper for publication has been based only on the merits of the work and the relevance to the journal.

Title, date and place of the conference

1st International Congress on Coastal Research (ICCR 2020),
July 6-9, 2021,
Al Hoceima, Morocco

Proceedings editor(s):

El Moussaoui Said, Abdelmalek Esaadi University, Morocco
El Talibi Hajar, Abdelmalek Esaadi University, Morocco
Cherkaoui Dekkaki Hinde, Abdelmalek Esaadi University, Morocco
Etebaai Issam, Abdelmalek Esaadi University, Morocco

Date and editor's signature

https://doi.org/10.1051/e3sconf/202129800001

Preface

The 1st International Congress on Coastal Research (ICCR 2020), held in Al Hoceima, Morocco (06-09 July 2021), was initially scheduled for 2020 but has been postponed due to the global COVID 19 pandemic. This first edition of a long series was attended by more than 150 scientists worldwide. A hundred communications (Conferences, oral communications, and posters) were presented, either in person or virtually. These were dispatched within six sessions covering the majority of Coastal topics.

The ICCR 2020 was a special occasion for those who develop coastal sciences in education, research, and professional development. It was an occasion to meet, listen, discuss, share information, and plan for the future. Indeed, a congress is an opportunity to discuss key points of scientific and professional policies and strategies, helping us to define how we should encourage responsible conduct in our political decisions, to promote marines and coastal sciences at the Moroccan level, and also establish personal contacts with colleagues from other parts of the world. The theme of this first edition was "Natural And Technological Risk Assessment Within Integrated Coastal Zone Management".

This Proceeding is a compilation of the most relevant communications presented in the congress. It provides an excellent overview of current coastal research, including but not limited to relevant aspects of geological sciences, biology (ecology and coastal marine ecosystems), geomorphology (physical geography), climate, littoral oceanography, coastal hydrogeology, environmental (resource) management, engineering, and remote sensing, is a reference platform for researchers, practitioners, and educators. It presents and discusses the most recent innovations, trends, concerns, and practical challenges encountered and solutions adopted in coastal environment sciences.

I want to express my deep gratitude to the Steering Committee and TPC members for dedicating their time to the organization of the conference. On behalf of the Steering Committee, I wish to thank Abdelmalek Essaadi University (Tetouan) and the Faculty of Sciences and Techniques of Al Hoceima for their support. We are also all very grateful to the National Center for Scientific and Technical Research, the Hassan II Academy of Sciences and Technologies, AGIR Association, Africa Geo-Services, Moroccan Committee of the International Association of Hydrogeologists, and Loukkos Hydraulic Basin Agency, that have kindly provided financial and logistical support in a challenging economic period.

I sincerely hope that the ICCR turned out to be a forum for excellent discussions that enabled new ideas to come about, promoting collaborative research. Finally, we wish to see you among us in the second edition (Tangier 2023) in better conditions.

Said El Moussaoui, Ph.D.
Publication-chair
1st International Congress on Coastal Research
(ICCR 2020)

https://doi.org/10.1051/e3sconf/202129800001

ICCR 2020 1st International Congress on Coastal Research
6-9 July, 2021 Al Hoceima, Morocco

Steering Committee

Pr. Hinde CHERKAOUI DEKKAKI
Vice President of Scientific Research & Cooperation
Abdelmalek Essaadi University, Morocco

Pr. Mohammed BAKKALI
Dean of the Faculty of Sciences and Techniques of Al Hoceima
Abdelmalek Essaadi University, Morocco

Pr. Hajar EL TALIBI
Faculty of Sciences and Techniques of Al-Hoceima
Abdelmalek Essaadi University, Morocco

Pr. Said EL MOUSSAOUI
Abdelmalek Essaadi University, Morocco

Pr. Issam ETEBAAI
Faculty of Sciences and Techniques of Al-Hoceima
Abdelmalek Essaadi University, Morocco

Pr. Lahcen BENAABIDATE
Faculty of Sciences and Techniques of Fez
Sidi Mohamed Ben Abdellah University, Morocco

Mr. Nour Eddine EL KADDOURI
Faculty of Legal, Economic and Social Sciences
Abdelmalek Essaadi University, Morocco

Pr. Hassan AMHAMDI
Faculty of Sciences and Techniques of Al-Hoceima
Abdelmalek Essaadi University, Morocco

Pr. Najat AMARJOUF
Faculty of Sciences and Techniques of Al-Hoceima
Abdelmalek Essaadi University, Morocco

Pr. Mohammed HASSANI ZERROUK
Faculty of Sciences and Techniques of Al-Hoceima
Abdelmalek Essaadi University, Morocco

Pr. Yassine EZZOUBI
Faculty of Sciences and Techniques of Al-Hoceima
Abdelmalek Essaadi University, Morocco

Mr. Houssine NIBANI
AGIR Association,
Al Hoceima – Morocco

Technical Program Committee

A

Pr. ABOUMARIA Khadija
Abdelmalek Essaadi University
Tangier, Morocco

Pr. AMARJOUF Najat
Abdelmalek Essaadi University
Al Hoceima, Morocco

Pr. AMHAMDI Hassan
Abdelmalek Essaadi University
Al Hoceima Morocco

Pr. AMRAOUI Fouad
Hassan II University
Casablanca, Morocco

B

Pr. BENAABIDATE Lahcen
Sidi Mohamed Ben Abdellah University
Fez, Morocco

Dr. BENAZZOUZ Omar
University of Aveiro
Aveiro, Portugal

Pr. BRIAK Hamza
Mohammed VI Polytechnic University
Benguerir, Morocco

C

Pr. CHERKKAOUI DEKKAKI Hinde
Abdelmalek Essaadi University
Al Hoceima, Morocco

D

Pr. DEBIECHE Taha Hocine
Benyahia University
Jijel, Algeria

E

Ech-charrat Mohammed Rida
Abdelmalek Essaadi University
Tetouan, Morocco

Pr. EL AMRANI Namira
Hassan the first University
Settat, Morocco

Pr. EL HAJRI Aimad
Cadi Ayyad University
Marrakech, Morocco

Pr. EL MOUSSAOUI Said
Abdelmalek Essaadi University
Al Hoceima, Morocco

Pr. EL OUARGHI Hossain
Abdelmalek Essaadi University
Al Hoceima, Morocco

Pr. EL TALIBI Hajar
Abdelmalek Essaadi University
Al Hoceima, Morocco

Pr. ETEBAAI Issam
Abdelmalek Essaadi University
Al Hoceima, Morocco

Pr. EZZOUBI Yassine
Abdelmalek Essaadi University
Al Hoceima, Morocco

F

Pr. FAOUZI Mohamed
Moahmmed V University
Rabat, Morocco

H

Pr. HAIDA Souad
Ibn-Tofail University
Kenitra, Morocco

Pr. HAMED Younes
University of Gafsa
Gafsa, Tunisia

Dr. HOUSSA Rachida
National Fisheries Research Institute
Casablanca, Morocco

I

Pr. ISAAD Jalal
Abdelmalek Essaadi University
Al Hoceima, Morocco

J

Dr. JARRAYA HORRICHE Faten
Water Research and Technology Center
Soliman, Tunisia

Pr. JOUNAID Halima
Faculty of Sciences of Ain Chok
Hassan II University
Casablanca, Morocco

L

Dr. LALILTI Anas
Abdelmalek Essaadi University
Tangier, Morocco

M

Pr. MHAMMDI Nadia
Mohammed V University
Rabat, Morocco

Pr. MOURABIT Taoufik
Abdelmalek Essaadi University
Tangier, Morocco

O

Dr. OMIRA Rachid
Instituto Português do Mar e da Atmosfera
Lisboa, Portugal

R

Pr. ROSSI Abdelhamid
Abdelmalek Essaadi University
Tangier, Morocco

T

Pr. TAHAYT Abdelilah
Abdelmalek Essaadi University
Tangier, Morocco

https://doi.org/10.1051/e3sconf/202129801001

Flood risk modeling in the watershed of Charâa stream, Province of Berkane (North-Eastern Morocco)

Hafssa Afif[1,*], *Mostafa* Oujidi[1], *Najat* Amarjouf[2], *Smahane* Ramdani[1], and *Abdellah* Azougay[1]

[1]Applied Geosciences Laboratory "LGA", Department of Geology, Faculty of Sciences Oujda - FSO, Mohammed First University, Oujda.
[2]Research and Development in Engineering Sciences Laboratory "LRDSI", Department of Earth and Environmental Sciences, FSTH, Abdelmalek Essaadi University, Tetouan.

Abstract. The knowledge on floods in Morocco is sporadic and scattered between historical writings and natural events following each other in time and space. In this context, the identification of areas likely to be affected by floods is essential for better management of this risk. The present study aims to model the floods at the level of the two sides of the Charâa stream (the side of Berkane and Sidi Slimane Charâ). The methodology followed in this modeling consists of determining the geomorphology of the watershed and studying its hydrology. 2D hydraulic modeling by Iber software makes it possible to follow the Spatio-temporal evolution of the characteristics of a flood. In addition, it allows us to map the hazard. Furthermore, to assess vulnerability to flooding, we referred to land use (physical vulnerability) to study the behavior of runoff water in the face of socioeconomic issues such as buildings, facilities, and the road network. The results showed that: (1) - A predominance of high-intensity flood hazard in the floodplain; (2) - Land with medium and high vulnerability successively constitutes 46.50% and 31.10% of the total studied watershed area.

1 Introduction

In Morocco, the floods of recent years have been catastrophic in many sites. Several urban areas are currently facing a severe urban management problem due to the flood risks caused by the excessive growth of these urban entities, without adequate sanitation and protection infrastructure[1].

In the Mediterranean region, past floods and forecasts of an increase in their frequency make it necessary for local planners and decision-makers to consider the risk of flooding in territorial management.

In Berkane, urban development in flood zones is one of the main factors aggravating vulnerability. In addition, developments (activities, infrastructure networks) modify flow conditions (waterproofing and runoff) while reducing the flood expansion. On watercourses, the developments (bridge, riprap) and the chronic lack of maintenance on riparian residents aggravate the risk of flooding. Indeed, the occupation of flood areas by buildings and materials sensitive to water can generate, in the event of flooding, the transport and deposition of undesirable products. These products can form obstacles in the stream bed. If they are broken, this can lead to the sudden flooding of downstream areas.

Effective watershed management and the subsequent utilization of water resources have become increasingly important in recent decades, although such management can be complex in rural areas that lack meteorological data. Scientists and engineers around the world have adopted hydrological and hydraulic models for modern-day flood forecasting. Hydrological models can estimate flow rates at outlets for various return periods, while hydraulic models can dynamically simulate flow regimes in rivers and flooded valleys.[2]

The present study focuses on the difficulty of flood risk generated by the conjunction of two components of a different nature, hazard and vulnerability. The essential steps in determining this hazard and vulnerability were estimating the maximum flood discharge using regional methods and 2D hydraulic modeling. With this study, we try to present universal methods that could be applied to other similar cities at the local spatial scale. [3]

2 General Contexts

2.1 Geographic Context

The province of Berkane is located in the extreme North-East of the Kingdom and the Oriental region. It is bordered to the north by the Mediterranean Sea, to the east by the Morocco-Algerian border, and the prefecture of Oujda-Angad, to the west by Nador, and the south by the province of Taourirt. Covering a total area of 1985 Km2, the province of Berkane represents 2.2% of the region's total area. [3]

* Corresponding author: afif.hafssa@ump.ac.ma

The diversification of its natural environment, the richness of its cultural heritage, and its privileged geographical location give the province particular assets and potential for agricultural and tourist activities.[4]

2.2 Geological Context

The province of Berkane is characterized by contrasting reliefs from North to South: the Triffa Plain, the Béni-Znassène mountain range, and the Angad Plain. [4]

2.3 Hydrographic Context

A branched hydrographic network characterizes this territory. The Moulouya river is considered the main watercourse of the province and the entire region and other less essential watercourses [5].

This study is based on the hydrological regime of the Charâa Stream. The latter drains three deep and narrow valleys in the Beni-znassène upstream of Berkane: Zegzel gorges, Ouartass and Beni-Ouaklane. However, it is above all the Zegzel Stream that ensures its durability due to sources of Liasic origin. The Cherrâa stream flows on a straight path of 12 km (the shortest distance between Moulouya and Berkane) to reach Moulouya. Downstream of Berkane, over a distance of about 4 km, it feeds the water table, then beyond this section, it drains the water table to Moulouya.

2.4 Climatological Context

The area is characterized by a Mediterranean climate (Köppen classification; Csa); Berkane has an average temperature of 18.2°C with a maximum average of 32.4°C and a minimum of 6°C. The total annual rainfall varies between 350 mm and 530 mm and is characterized by a Spatio-temporal irregularity. [7].

3 Materials and Methods

Risk can be interpreted as the confrontation of hazard and vulnerability in a specific place and at a specific time; hazard being considered as the probability of occurrence of a natural, potentially dangerous, and potentially damaging phenomenon to property and people; and vulnerability as the magnitude of damage or loss resulting from the occurrence of a given natural phenomenon. [9]

This study is carried out according to an adapted scientific protocol using computer tools for data processing:
- The Global Mapper software: a tool for downloading the satellite image of the study area (Berkane region).
- ArcGis 10.6.1 software: GIS tool for digital terrain modeling and land cover digitization.
- The application of the "SCS" method for the quantification of the peak flow.
- The Iber software: a tool for data simulation and 2D modeling of the hazard.

The study is carried out in the following steps:
- DEM preparation and satellite imagery.
- The study of the sub-watershed of Charâa Stream, which includes the cities of Berkane and Sidi Sliman Charâa fed by Ouertass Stream and ZagZel Stream as two main drains by delimiting the sub-watershed, by classifying the rivers on ArcGis by the hydrological modeling tools and by transforming rainfall into flow by the SCS method after deducing its morphological characteristics:
-The surface area of the basin is measured in hectares or square kilometers.
- Length is measured in meters or kilometers.
- The slope is measured as a percentage.
- Concentration time is when it takes for a drop of water to travel the length of the basin. This is the minimum amount of time a rainfall must last for the entire pond to react. It is expressed in hours.
- The production of a land use map that allows an understanding of its organization and spatial dynamics and that is produced from satellite images.
- Drawing up a map of the territory's vulnerability to flooding, which shows the sensitivity of the territory to a hazard and which is based on its physical vulnerability (the resistance of buildings, installations, networks, and infrastructures).

The hazard map development identifies areas where flooding is likely to occur, to a greater or lesser extent and with a greater or lesser frequency. It is based on combining the following two concepts: the velocity of the flood and the depth of water.

4 Results and Discussion

4.1 Hydrology of the sub-basin

The following figure (Fig.1) shows the delimitation of the sub-watershed of Charâa Stream characterized by:

Table 1. Characteristics of Charâa Stream

Surface (km^2)	Lengt h (km)	Perimeter (km)	Slope (m/m)	KG	Tc (h)
197.299	29.988	83.88	0.0442	1.672	5.1 6

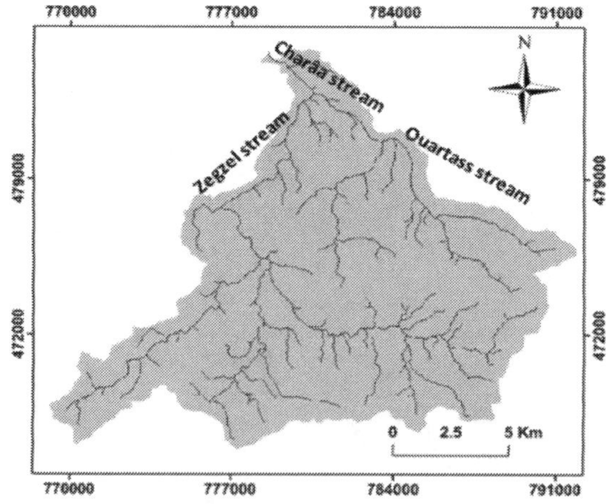

Fig 1. Map of the Charâa Stream sub-**watershed**

Applying the SCS method, we obtain the following hydrograph (Fig.2), which presents the instantaneous debits in m3 /s, with time in hours with a base time of 14h, a maximum debit of 580.1 m3/s.

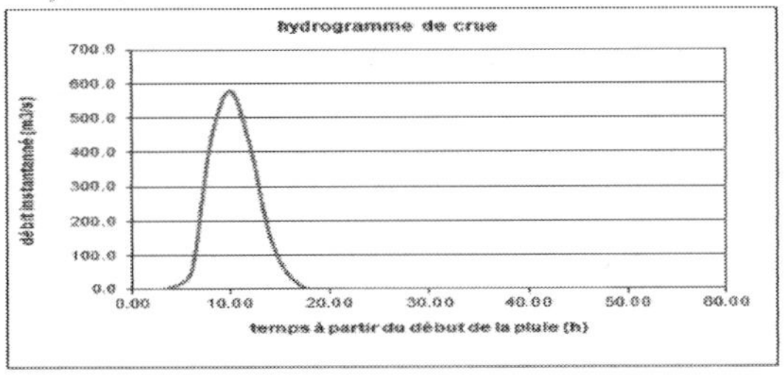

Fig.2. The flood hydrograph (SCS method)

4.2 Land Use

The land use map of the study area illustrates the different types of urban, rural, mountainous environments, etc., based on the classification of the Iber 2.5 software (bare soil, dense vegetation, sparse vegetation, industrial zone, infrastructure, residential zone, forest, river, unclassified zone, urban vegetation).

For our problem, this land use (Fig.3) will be used to study the behavior of runoff water as a function of the porosity of each type of soil and to define its vulnerability.

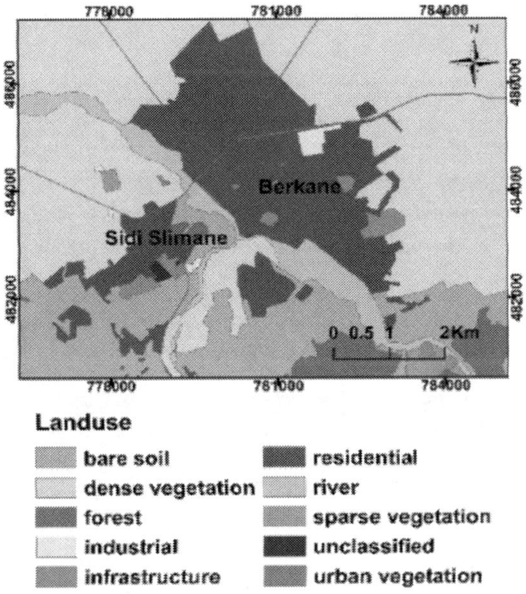

Fig.3. land use map of the study area

4.3 Vulnerability

The physical vulnerability for each type of land use takes values between 0 and 1 and between low and high (Table.2), which depends on the importance of this occupation about the presence of residents and goods and infrastructure.

The results obtained by mapping on the ArcGis tool (Fig.7) show that the study area is characterized by a high vulnerability occupying 31.10% of the studied area and a medium vulnerability occupying 46.50% of the said area. This high vulnerability can increase the risk of flooding.

Table 2. Vulnerability Indicator

Land use	Vulnerability indicator	
	from 0 to 1	in terms
Bare soil/ Sparse vegetation/ forest/ Unclassified /Urban vegetation	0	Low
Dense vegetation	0.5	Med
Industrial/ infrastructure/ Residential / rivière	1	High

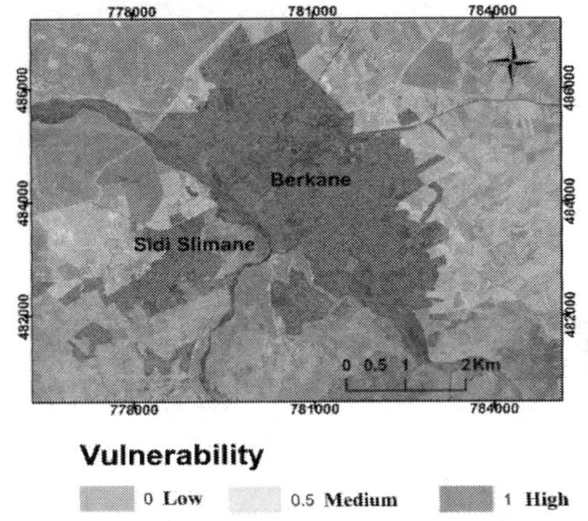

Fig.4. Vulnerability map for Berkane and Sidi Sliman Charâa

4.4 Hazard

The hazard map is the result of the hazard study, and the objective is to assess the hazard intensity at any point in the study area.

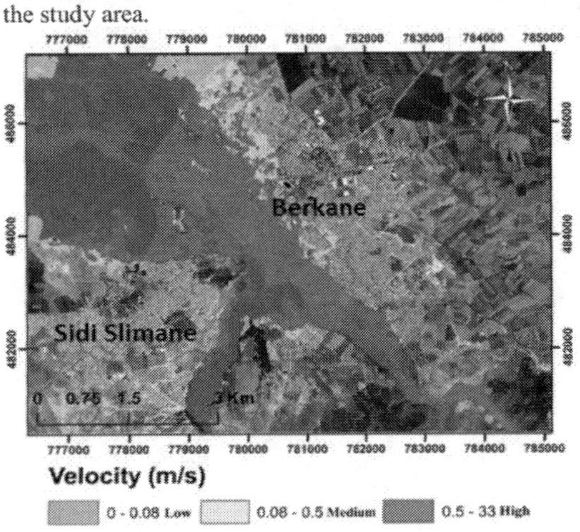

Fig.5. Flow velocity map of Charâa Stream

https://doi.org/10.1051/e3sconf/202129801001

The hazard represents flooding as a physical, natural, and uncontrollable phenomenon, regardless of its potential effects on the environment and human activities. Hazard depends on the hydrological behavior of the catchment area and the hydraulic functioning of the river system.

Our first factor in determining hazard is the flow velocity (Fig.5).

Our second factor in determining hazards is water depth (Fig 6).

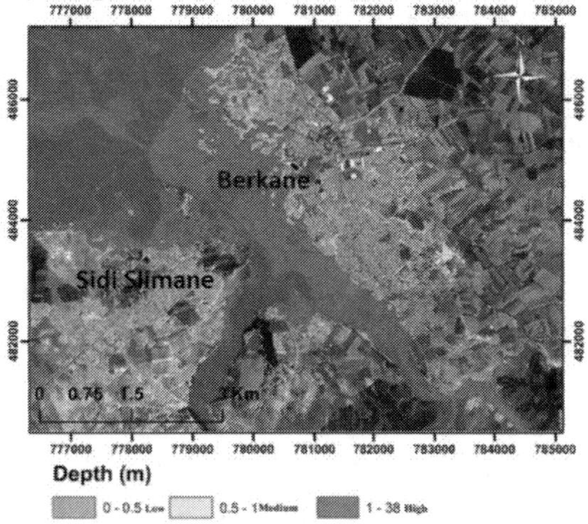

Fig.6. Water depth map of Charâa Stream

After multiplying the two flow velocity maps (Fig. 5) of the flood and the water depth (Fig. 6), we obtain the hazard map using the ArcGIS tool (Fig. 7):

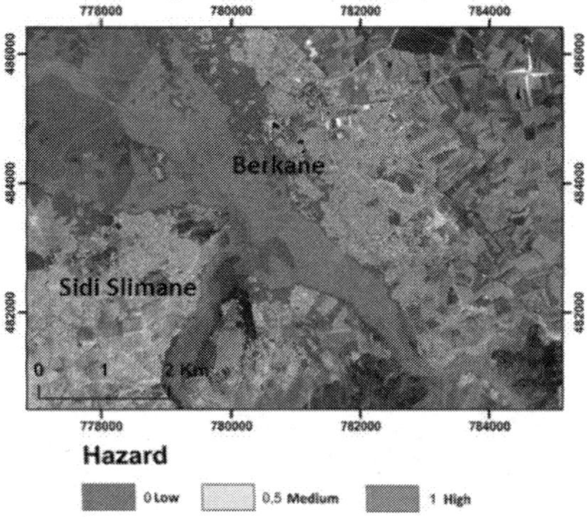

Fig.7. Hazard map for Berkane and Sidi Sliman Charâa

Our study area is characterized by a high flow velocity exceeding 0.5m/s and a significant depth exceeding 1m. Combining these two factors gives an extreme hazard over a large area at the level of the Berkane city and Sidi Slimane Charâa. This result gives us an idea of the danger under which the region is exposed in the case of a centennial flood.

5 Conclusions

The study of hazard and vulnerability on the site of Berkane and Sidi Sliman Charâa highlights the physical characteristics of the floods and their stakes. Urban development aggravates the flooding phenomenon of Stream Charâa, which exposes a significant population facing this risk and the infrastructures of the area.

The knowledge of hydrological mechanisms and human occupation of the land is insufficient to solve flooding in Berkane and Sidi Slimane Charâa. On the other hand, it is only a first step towards better risk management, particularly by proposing competent authorities cartographic tools for decision support.

Acknowledgments

This study was carried out thanks to the excellence grant from the National Center for Scientific and Technical Research (CNRST)

References

1. D. Elhafid. B. Akdim. Echelle Spatiale Et Genese Des Risques Hydrologiques En Milieu Urbain Et Peri-Urbain : Cas De La Ville D ' Oujda. 10 (2004).
2. N. Amarjouf. M. Oujidi. D. Meijer. W. Klop. ISH Hydraul Eng. Evaluating ephemeral watercourse discharges by hydrologic-hydraulic modeling: application to the Kert River basin, Morocco. 11. (2019).
3. Z. Naiji. M. Oujidi. N. Amarjouf. and H. Rezqi. Application of Two-Dimensional Hydraulic Modeling in Flood Risk Mapping. A Case of the Urban Area of Zaio, Morocco. 22. (2019)
4. Ministère de l'Aménagement du Territoire National, de l'Urbanisme, de l'Habitat et de la Politique de la Ville. Appel d'Offres Ouvert N ° 25 / 2018 ETUDES RELATIVES A L'ELABORATION DES CARTES D'APTITUDE A L'URBANISATION LOT N ° 1: RELATIF A LA PROVINCE DE BERKANE. 33. (2018).
5. MINISTETRE DE L'INTERIEUR/ PROVINCE DE BERKANE. La province de Berkane en chiffres 1. 33. (2006).
6. S. MILHI. A. SBAI. K. BENRBIA. Les enjeux sociaux du tourisme dans la province de Berkane. 21. 113–134. (2018)
7. N. Bechler-Carmaux. M. Mietton. M. Lamotte. Ann. Georgr . Le risque d'inondation fluviale à Niamey (Niger). Aléa, vulnérabilité et cartographie. 11. 176–187. (2020).

Earthquake-induced liquefaction in the coastal zone, Case of Martil city, Morocco

Mohamed El Hilali[1,*], *Youssef* Timoulali[1], *Taj* Benyounes[1], *Mohamed* Ahniche[2], *Rabie* El Bardai[2], and *Sahel* Yattara[3]

[1]Earth Sciences Department, Scientific Institute, Mohammed V University in Rabat, Morocco
[2]GeoTechMed Laboratory, Tetouan Park: lot n°11, Tetouan, Morocco
[3]Abdelmalk Essaadi University, Geology Department, Faculty of Sciences, Tetouan, Morocco

Abstract. According to historical documents and Moroccan earthquakes catalogs, the coastal zone has suffered in the past from several earthquakes. Understanding how sedimentary basins respond to seismic-wave energy generated by earthquake events is a significant concern for seismic-hazard estimation and risk analysis. The main goal of this study is to determine the distribution of the natural frequency value (F), the amplification factor value (A), and the soil vulnerability index (Kg) were carried out as an indicator for liquefaction potential sites in the Martil city based on the microtremor measurements. Liquefaction assessment was done at 96 stations using the HVSR approach provided by Nakamura (1989). According to the analysis results, the predominant frequency values range from about 0.31 to 5.63 Hz, and the amplification factor values range from 3 to 15. Based on these parameters, the vulnerability index Kg is determined, which can be used as a parameter in calculating the liquefaction potential of an area. This study shows supporting evidence for the first time that the HVSR of microtremors can be an excellent alternative indicator for an area's potential for liquefaction.

1 Introduction

Over the previous decades, many earthquakes have manifested worldwide, leaving behind it many injuries, deaths, and many buildings damaged. Experience shows that the catastrophic effects of earthquakes have continued for several years. Therefore, it is necessary to guarantee the safety of inhabitants through several correct programs designed to provide a secure living space.

Based on the geophysical and geological data (Fig. 1B), the plate boundary between Africa and Europe is well defined from west to east respectively from the Azores in the Atlantic to the west of the Strait of Gibraltar is punctuated by earthquakes of magnitude greater than 6 (Lisbon earthquake of 1755: M = 8. 7; 1941: Ms = 8.2; 1969: Ms = 7.3; 1975: Ms = 6.7; 2007: Mw = 6.1) tracing the activity of a transform fault. Beyond the Strait of Gibraltar, towards the east (the Alboran Sea and surrounding areas), the seismicity is low to medium and diffuse, characteristic of a continental collision zone [1].

Martil is a Mediterranean coastal plain located east of Tetouan city, specifically 10 km away, at the northwest edge of Morocco (Fig.1A). The results of the demographic projections of the prefecture M'diq Fnideq show that the population of the municipality of Martil would be 119 128 inhabitants by 2030 against 63 853 in 2014 (HCP 2014-2030), covering a surface of 87 km².

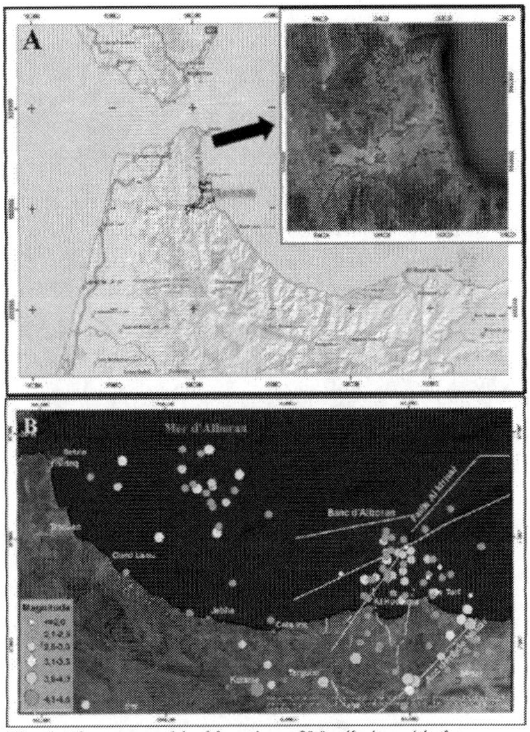

Fig. 1. The geographical location of Martil city with the seismicity of the Rif and the Alboran Sea (after Cherkaoui 2019). https://mtcherkaoui.wixsite.com/site

* Corresponding author: elhlalimohamed96@gmail.com

 https://doi.org/10.1051/e3sconf/202129801002

According to these data, the population of Martil city has been increasing rapidly in the last decades and the future. For this reason, the present study is realized in this area to mitigate the seismic risk by assessing the soil liquefaction phenomenon as the main result caused by an earthquake.

Recently, the H/V technique [2] is becoming more popular with its data collection facilities due to it being easy to use anytime and anywhere. In this study, we analyzed the site response using the HVSR method in Martil city. Data were preprocessed using Grilla software 2014 (Fig. 2) to estimate the fundamental frequency (F), amplification (A), and the vulnerability index "Kg value" was calculated for each observation point as a means to determine the extent of liquefaction.

The present study mainly aims to show how the microtremor measurements can play a significant role in a liquefaction assessment [3] and show the possible impact of an earthquake in Martil city.

2 Methodology

The Microtremor method (MHVSR - Microtremor Horizontal to Vertical Spectral Ratio) is a technique that calculates the ratio of the horizontal to vertical Fourier (amplitude) spectra derived from microtremors recorded by a three-component sensor [4]. There is empirical evidence that the maximum or the peak in the MHVSR spectral ratio appears at the frequency, which equals the resonant frequency of sediments at a specific site [5]. Recently, to assess potential hazards associated with ground failure or liquefaction, the vulnerability index (or Kg value) derived from microtremor data may be evaluated. Among the various approaches to microtremor study, the H/V spectral ratio technique introduced by Nakamura (1989) [2] was chosen for this investigation due to ease of application. This technique has been described in many papers.

By employing the H/V ratio, we determined the predominant frequency (F) and the site's amplification factor (A). Nakamura also proposed the vulnerability index "Kg value" as a means to determine the extent of liquefaction [6].

For this purpose, he compared the results of the vulnerability index and the damage points in the Kobe earthquake and concluded that Kg is a suitable index for determining the vulnerable site points, such that each point with a higher Kg value is more damage potential. The Kg value is derived from the strain of ground structures. It can be defined as:

$$Kg = A^2/F \quad (1)$$

With A is the amplification factor and F is the fundamental frequency.

After Nakamura, some scientists have used microtremor to liquefaction evaluation [7-8].

3 Data collection and results

This study used the microtremor measurements of 96 sites in Martil city during the period 2018 and 2019.

Microtremor measurements were taken at 96 points in Martil city. For all stations, the microtremors were measured, and the HVSR spectral ratios were calculated by the procedure used in the Grilla software packages prepared by the SESAME group.

Figure 2 shows the H/V spectral ratio for some stations. According to the analysis, the fundamental frequency (F) varies from 0.31 to 5.63 Hz within the study area. Also, the amplification factor (A) varies from 3 to 15 within the study area. Figure 3 shows the fundamental frequency and amplification factor in the study area. As shown in figure 3 for some recording stations, fundamental frequency and amplification factors have not been presented. Since the factories and highways have surrounded these stations in the city, accurate microtremor recording was impossible all day long [9], [10].

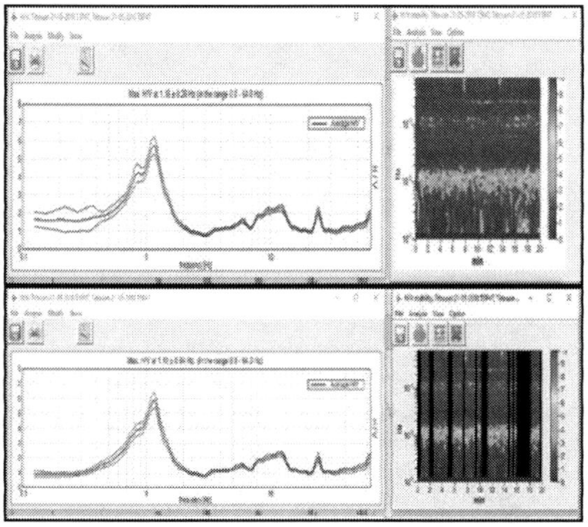

Fig. 2. H/V spectral ratio

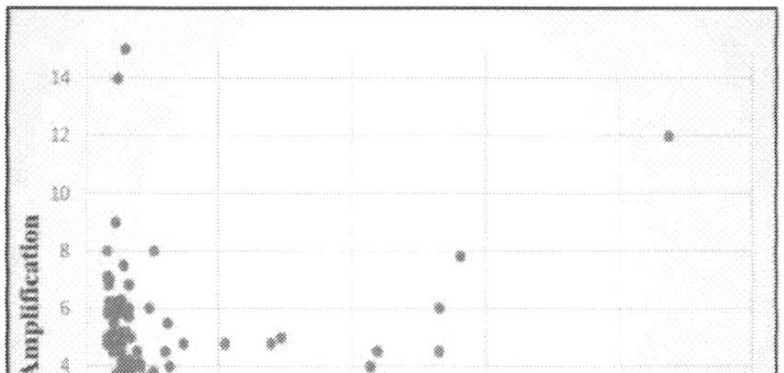

Fig. 3. representation of the fundamental frequencies with their amplitude in the different stations recorded in the city of Martil.

By using the results of figure 3, the Kg values are calculated for different stations. These results are

 https://doi.org/10.1051/e3sconf/202129801002

indicated in figure 4. By reviewing the scientific literature, it has been concluded that the higher the Kg value, the more liquefaction potential will be. In fact, the Kg value has a qualitative meaning, for instance, the possibility of liquefaction occurrence.

As presented in figure 4, Martil City has a seismic vulnerability index going more than 30. Several areas with a seismic vulnerability index of about 0-20 were found on some parts of the coastal area, especially at the north parts of Martil City. It is observed that in all the stations with Kg values higher than 30, liquefaction phenomena occurred. Consequently, the value of Kg=30 can be considered a threshold value for these 96 stations, in that for all the stations of this region with a Kg value higher than 30, the occurrence of liquefaction phenomena is possible in stations with lesser values; this possibility does not exist.

For the areas susceptible to the phenomenon of soil liquefaction with kg>30, as indicated in figure 4, having important amplitudes and a lithology essentially of ancient and recent alluvium typically of quaternary age resting on a substratum of Pliocene age. These alluviums are deposited by two rivers, which cross it transversally, the river Alila in the north of Martil and the river Martil that extends for 60 km. However, other areas (kg<30) are less susceptible to this phenomenon than the previous one and have different lithologies of marly, sandy type, etc.

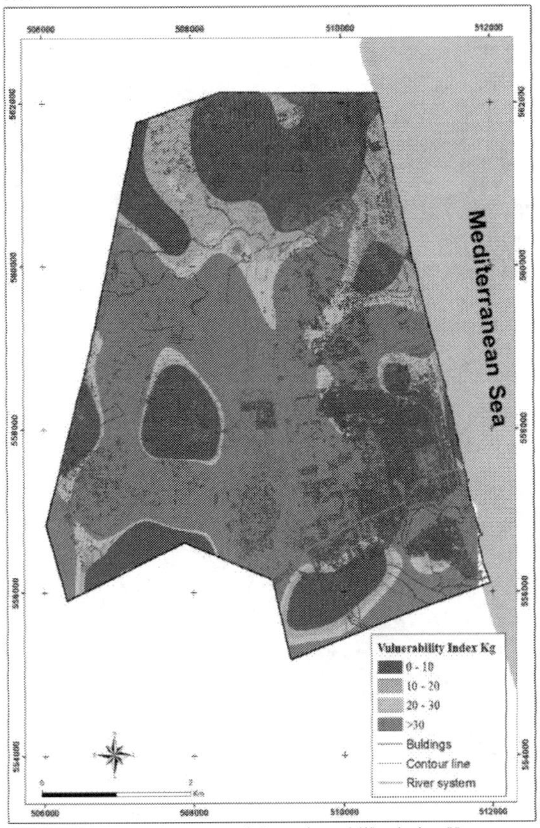

Fig. 4. Microzonation map of the vulnerability index Kg

4 Conclusion

Due to the geological structure and the nature of the soils of the province of Tetouan and the prefecture of Fnideq-M'diq, site effects (local amplification phenomenon) may likely be observed in the case of seismic solicitation. It would be wise to carry out the necessary and adequate measures to evaluate such an effect and, consequently, anticipate any policy of mitigation of the seismic risk.

This paper presented seismic hazard mitigation based on the vulnerability index indicator for the Spatial Plan of Martil City, Morocco. The geophysical characteristics, such as amplification (A), predominant frequency (F), and seismic vulnerability index (K_g), were presented. Generally, several areas in Martil City had larger amplification, especially in the places near the sea. It was indicated by the larger value of the fundamental frequency. The larger Kg values were found in the city center and the southern part of Martil City (Kg>30). It indicated that a significant seismic effect could occur in Martil City. Along this vulnerable area, the high density of people concentration existed. The finding obtained from this study could recommend the local government to consider the earthquake impact in Martil City, especially along in areas with a higher vulnerability index.

References

1. P. Tamponnier: Tectonic evolution of the alpine system in the mediterranean, punching and rigid-plastic erading. bulletin of the geological society of France (1977) S7-XIX (3): 437-460 (1977).
2. Y. Nakamura, "A method for dynamic characteristics estimation of subsurface using microtremor on the ground surface". Quarterly Report of RTRI, Vol. 30, pp. 25–33 (1989)
3. HC. Huang, YS. Tseng "Characteristics of soil liquefaction using H/V of microtremors in Yuan-Lin area, Taiwan", TAO, Vol. 13(3), pp. 325–338 (2002)
4. S. Molnar, J.F. Cassidy, S. Castellaro, C. Cornou, H. Crow, J.A. Hunter, S. Matsushima, F.J. Sanchez Sesma, and A. Yong : Application of microtremor horizontal-to-vertical spectral ratio (MHVSR) analysis for site characterization: State of the art. Surv. Geophys, 39, 613–631 (2018)
5. E. Field. and K. Jacob : The theoretical response of sedimentary layers to ambient seismic noise. Geophys. Res. Lett., 20, No. 24, 2925–2928 (1993)
6. Y. Nakamura "Real-time information systems for hazard mitigation", 10th World Conference in Earthquake Engineering, Spain, Madrid (1996)
7. S. Rezaei, AJ. Choobbasti, "Liquefaction assessment using microtremor measurement, conventional method and artificial neural network (Case study: Babol, Iran)", Frontiers of structural and civil

engineering, DOI 10.1007/s11709-014-0256-8 (2014)

8. J. Saita, Y. Nakamura, T. Sato, "Liquefaction caused by the 2011 off the Pacific coast of Tohoku earthquake and the result of the prior microtremor measurement", 9th International Conference on urban earthquake engineering, Tokyo, Japan, (2012)
9. AJ. Choobbasti, S. Rezaei, F. Farrokhzad, "Evaluation of site response characteristics using microtremors", Gradevinar, 65, pp.731-741 (2013)
10. AJ. Choobbasti, S. Rezaei, F. Farrokhzad, Heidarzadeh P. Azar "Evaluation of site response characteristics using nonlinear method (Case study: Babol, Iran)", Frontiers of structural and civil engineering, DOI 10.1007/s11709-014-0231 (2014)

Geological and geomorphic investigations on palaeo-landslide dammed the Tamourghout River (Middle Atlas, Morocco)

Jamal Abbach[1,*], *Hajar* El Talibi[1], *Said* El Moussaoui[1], *Hinde* Cherkaoui Dekkaki[1], and *Issam* Etebaai[1]

[1]Faculty of Sciences and Techniques of Al Hociema, Abdelmalek Essaadi University, Tetouan Morocco

Abstract. The Middle Atlas Mountains in Morocco are best known for their slope-active movements, where lithology, topographic relief, and seismotectonic are likely to cause slope instability problems. This study focused on geomorphological and geological investigations on palaeo-landslide blocked and dammed the Tamourghout river main bed (Taza, Morroco). The paper is essential due to the enormous proportions of the landslide and also because of the vast watershed lock-up by the dam (Tamda lake) and the recognition that a Tamourghout river has been and can be obstructed by a permanent landslide

1 Introduction

A great diversity of natural lakes characterizes the Atlas domains of Morocco. They are typical and educational examples in the reconstruction of the geological history of the Atlas Mountains. Their genesis depends on their origins which are generally caused by natural phenomena.

Lake Tamda, suspended between the altitudes of the northern Middle Atlas, is located 80 km southeast of Taza in the foot of Boublane at 1500m altitude; it is a typical and educational example of a natural lake landslide origin. Its natural dam was created by sliding a gigantic carbonate body, blocking the hydrographic network and giving rise to a large water reservoir with a length of more than 4 km and an average width of more than 200 meters.

2 Study area and geological setting

The study area is located in southeast Taza in the Magharwa region (Fig. 1), belonging to the folded part of the Middle Atlas chain (Fig. 2), near the Central Middle Atlas Fault (ACMA) boundary.

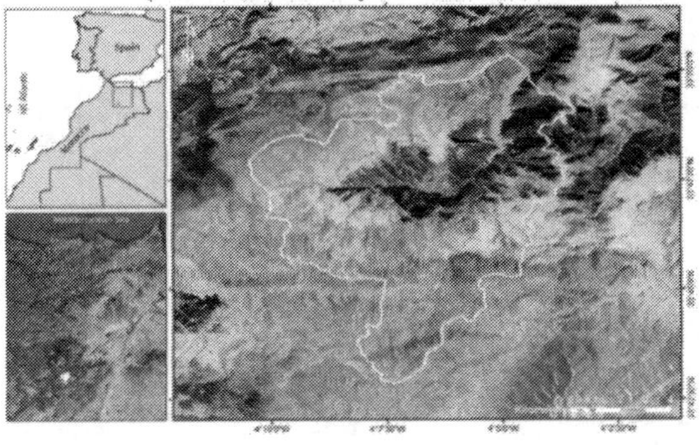

Fig. 1. The general setting of the study area with the location of lake Tamda

It includes two large mountain ranges, Tazekka hill in the west and Bou Ibalne in the East. The tectonic structure corresponds to large synclinal basins (Tamatroucht) with axes parallel to the chain Nar narrow anticline wrinkles sometimes intrude with gabbroic rocks [1-4].

The geological bedrock consists of Paleozoic, partly Carboniferous basement in exposed the Tazekka hill and covered by thick carbonates, clays, and basalts units mainly Mesozoic-Quaternary [4-8].

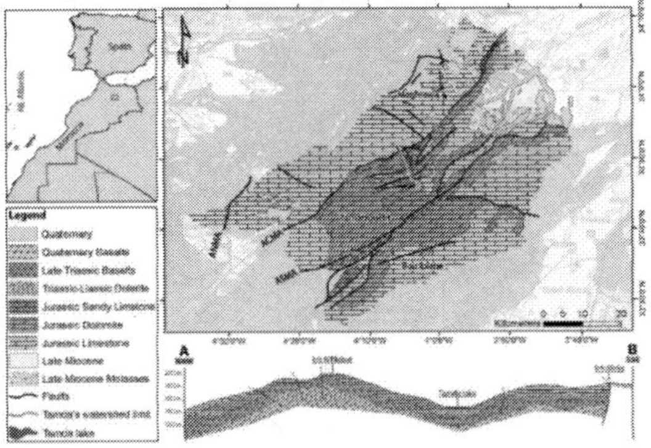

Fig. 2. Simplified geological map and geological section indicating sediments exposed in the study area (modified after [13])

3 Climate setting

In the Middle Atlas and surrounding areas, the climate is warm and temperate. The study area is characterized by large spatial variations of rains influenced by the windward effect, elevation, and aspect. The Fès-Taza corridor is characterized by a rainy winter rainfall regime, extending from November to April, and a relatively dry to arid summer, going from June to September. The annual rainfall average varies between 500 mm in the west (Sebou-Inaouène confluence) and 700 mm east of Taza

* Corresponding author: jamalabbach21@gmail.com

(Table1). Nevertheless, the mountainous areas bordering the basin are much more watered: up to 1500 mm in the southeast, on the peaks of the Djebel Tazzeka, and a little more than 1000 mm on the high northern slopes. However, the interannual precipitation variability is quite large, but with a total number of rainy days consistently low, 60 to 70 days on average per year. The snowmelt is also essential on high cliffs. It feeds groundwater supply and promotes limestones dissolution and the development of karst [9].

Table 1. Average annual rainfall and temperature between 1980 and 2019 (Data from National Forestry Research Center (CNRF)

Station	Latitude (N)	Longitude (W)	Altitude (m)	Rain (mm/an)	T (°C)
Bab Boudinar	34°04'	4°07'	1570	1220	19
Tazekka	34°11'	4°12'	1380	851.4	18
Taza	34°13'	4°01'	500	524.5	23
Bou Iblane	34°14'	4°02	2500	678.6	18
Tamatroucht	34°14'	4°01'	2450	678.4	17

4.Methods

The proof for the landslide dam has been assembled from field investigation, analysis of geological maps, and digital elevation models (ASTER-GDEM V2). Geomorphic parameters are collected from bibliographic archives analysis or estimated via cartographic and satellite imagery interpretation. We opted for a multidisciplinary approach to study the landslide: Slide morphology has been studied through satellite images. With the GIS system, we mapped and highlighted the elements of the slide. Direct field observations and photographing have been supplemented by field mapping. We focus on identifying specific characters of the karst features and, due to the specificities of the geological context, their characterizations as geo-trigger off a landslide.

5.Results and Discussion

5.1 Geomorphic characterization of Tamda watershed

The study area of Tamda (Fig. 3) has sub-vertical slopes where the slopes vary between 50 and 70%. The watershed from about 60km^2 is surrounded by a set of clips that exceed 2000m. A DEM with 30 m resolution was a great source to generate various topographic factors, which influence the landslide activities in this area. From this DEM, three thematic data layers were extracted, which are:

5.1.1 Hypsometry

The Hypsometry of the Tamda watershed is spread over several fractional surfaces as a function of altitude, as shown by the hypsometric map (Fig. 3A). The hypsometric classes between 1800 and 1900m mark 20.40% of the watershed with a surface that exceeds 16 km^2. The minimum Hypsometry is 1450 m which is the lake bottom point. It can be concluded that the studied terrain is characterized by considerable Hypsometry and reflects the mountainous dominance of the Tamda watershed

5.1.2 Slope

It consists of the maximum change in z-value from each cell. The slope is very steep throughout the area, especially in the dam's vicinity (Fig. 3B). It varies between 0 and 89° on the surface of the watershed. We have located in the field an extreme escarpment beyond the dams; the concentration of the maximum slope is very distinct in the northern part of the watershed, precisely near the slide body. We can also note on the map slopes areas with zero slopes that inevitably provide information on tectonic and/or karst depressions. The value of the slope that dominates the watershed is between 10° and 30°; thus, the studied area has a very particular inclination

5.1.3 Aspect

refers to the direction of the maximum values of those changes from each cell to its neighborhood in the output raster. The values of each cell in the output raster indicate the compass direction that the surface faces at that location. It is measured clockwise in degrees from 0° to 360° site ranges from 1°, which corresponds to the flat areas having no downslope, to 356° North. All the directions, as well as the corresponding degrees, are described in Figure 3C. Slope exposure in our study area is dominated mainly by North orientation 15.34% and Northeast 14.25%. The orientation of the slopes in the study area influences especially the degree of instability on the exposed slopes North and NNE for the landslides. Slopes of the direction South, SSE, SSW show especially an erosion dominated by the streaming.

5.2 Main features of the Tamda lake

The Tamda dams lake has a very long form (Fig. 3D). The length can reach five kilometers; it is fed essentially by the alluvium precipitations; it fills up during the cold season when rainfall is plentiful, the water level drops in summer and dries up during the persistence of warm seasons. It is of type "fill-and-spill". The dike is located at an altitude of 1500 m, and it is generally carbonated, covered by cedar; its height reaches 50 meters, its length and width measure 474 m and 426 m, respectively.

The watershed's hydrology is characterized by an endorheic flow that feeds the Tamda Lake, blocked by a very huge carbonated sliding body and constitutes a rigid and impervious dike. The primary hydrological resources that feed the water reservoir are the rainfall and snowfall inputs drained by the different rivers and their effluents towards the outlet. The drying up of the lake is mainly done by evaporation and infiltration, notably by continuing the extensions of the mid-atlas fault with an N10 direction. The watershed of this water reservoir is dispersed over an area in the vicinity of 78.69 km^2; its maximum length can reach 17, 16 km, its maximum width is 9.45 km, and its perimeter is approximately 51.93 km.

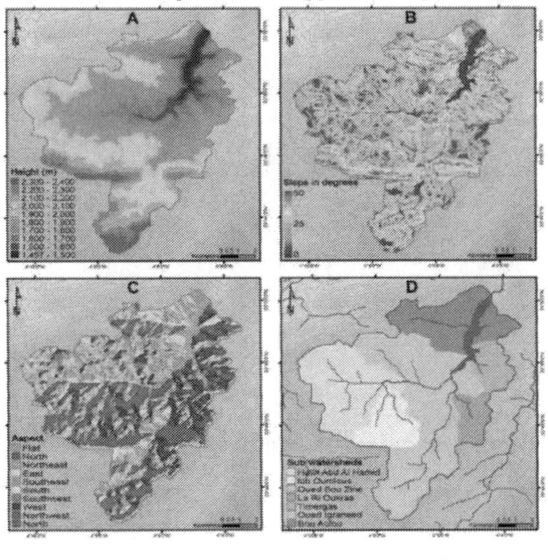

Fig. 3. the topographical characteristics of the Tamda watershed

5.3 Karstification features

The karst of the study area presents forms like rock cones, giant sinkholes, snow wells of the Jebel Bou Iblane, and skimmer of sinkholes rich in underground rivers. The respective roles of ancient climate conditions and the tectonic setting are difficult to specify. While, the genesis of the forms is replaced in the overall evolution of the Middle Atlas characterized by superimposition of rivers, the fluctuation of the Atlantic-Mediterranean watershed, and minor tectonic readjustments that guided karstification [9-11].

Fig. 4. A landscape presentation of examples of karst structures in the study area

Mounir et al. recognized in detail the karst forms of the central Middle Atlas near to Taza; this karst is essentially characterized by the variety of surface forms, lapidates, caves (Fig. 4), sinkholes, poljés, dry valleys, rock cones, and crypto karst developed in quaternary basaltic flows.

5.4 Landslide characteristic

Based on GIS analysis and geomorphic evidence, the landslide location and volume can be estimated. The length of the rock movement from the surface to the riverbed is 750 m with a width up to 1 km (Fig. 5). Given the slide oldness, it is hard to know the depth to the shear surface from the topography, and therefore any estimation of the slide volume is provisional. Studies of many of the significant landslides recommended that an expected average depth of 100 m is realistic. The volume of the landslide displaced on the river would then be at least a few hundred million m3 (750 to 1000 x 92).

The sliding body moved about 500 m and obstructed the Tamourghout bed, creating a landslide dam closely 1500 m high. The SE slope of jbel N'erkibat still shows the freshness of the main and lateral tearing scars. The sliding body crossed the bed of the Tamourghout river and ran on the opposite side of 300 m high, forming numerous thrusts and scarps of an anti-slope cliff. The main slope is serrated and sub-vertical (30°-60°), with a vertical height of 800 m. The deposits in the landslide dams are mainly composed of carbonate rocks (Fig. 5A-B), including limestone and dolomite Jurassic complex, sandy mudstone of the upper Miocene molasses units.

Fig. 5. The morphology of the main slide and its units.

https://doi.org/10.1051/e3sconf/202129801003

The field observations of the sliding surface allow us to classify the event in Rock/debris slide complex. This slide category is a large bulk of mostly dry rock debris caused by the collapse of a slope and run-up over a long distance [12]. Its speed can reach 10 m/s, and the volume is up to 100 m3. The moved distance of a rockslide frequently surpasses several kilometers, and the movement becomes evident by the run-up on the slope of the opposite side of the valley (Fig.6). On the study site, we were able to identify two types, the first one corresponds to active landslides (debris avalanches and debris flows) in the vicinity of the lake, and the second case corresponds to a sudden rock slide which caused the Ich Nrkibat phenomenal mass wich blocked Tamourghout river and create a large body of water. The sediment layers of the slide dams are heterogeneous and mixed with a wide variety of lithology and particle sizes. The main deposit area comprises fragments of fine sediments and debris mixed with large boulders and blocks with sizes ranging from 0.5 m to 5m (Fig. 5C-D). The middle and lower layers are better preserved, not strongly disaggregated, and formed of the almost intact rock body.

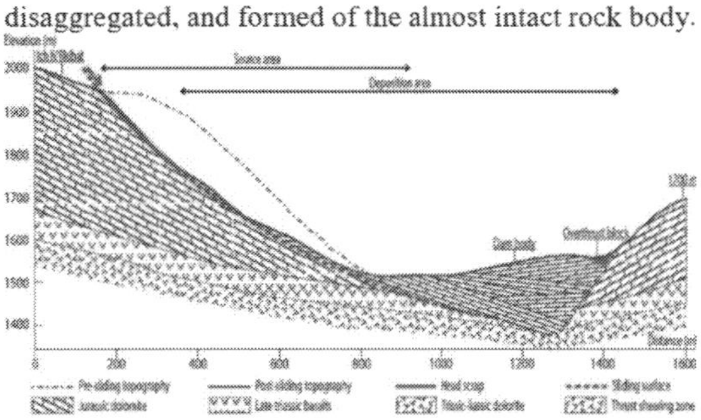

Fig.6 Geological section showing the lithology of the Tamda landslide

5.5 Discussion

Tamda is, without a doubt, a natural lake formed on the occasion of an immense catastrophic landslide, which blocked the deep valley of Tamourghout. This landslide is a slope instability process, which produces local geomorphic, hydrologic, and seismotectonic conditions. In fact, The Maghraoua syncline is one of the most fractured regions of the Middle Atlas. [13] has recognized two types of faults in the region: oriented faults NE and WNW oriented faults with expanding cracks. Thus, Tamda is located in a well-known tectonically active region with potential geological conditions for strong earthquakes [14]. The ACMA Fault is the main fault among the Tamourghout watershed, and it runs through this area along the valley. Based on historical seismic records, multiple historical seismic events with a magnitude of ≥ 6 since 1755 occurred in the study area [14, 15]. A seismic event triggered the landslide due to a replay of the central fault that probably accompanied the 11 May 1624 Fez earthquake. This quake is of the worst natural disasters in the history of Morocco and caused severe damage in the city of Fez, Taza, and its surroundings [15]. This interpretation is also based on many observations, including the substantial dip of the limestone layers towards the SE (30-70 °); the imbalance situation testified by the outcrop of the Triassic clay downstream due to the river erosion.

Generally, The Atlas Mountains are often prone to slope instabilities [16]. Since the Villafranchien, tectonic movements have caused large hills broken into bulky blocks testifying to a tectonic paroxysm. The role of seismic movements in this context of landslides is not to be demonstrated because they occurred at all landscape scales [17]. Thus, most of the seismic epicenters studied in the Middle Atlas coincide with faults alignments [15] [18], which attests to a strong correlation between the earthquake and geological structure. The same neotectonic activities were widely discussed throughout the North Middle Atlas accident.

Furthermore, this study revealed that the existing geomorphic evidence is in agreement with the results of previous studies [16] [19] [20] that is to say that earthquakes could have occurred in the Middle Atlas, along the Tamourghout River. Through examinations of the landslide aspect, we can deduce that the lake dammed by the paleo-landslides of Tamda is considered the result of a large avalanche of rocks triggered by an earthquake. The incidence of the Tamda landslide will be controlled by the combined effect of geological (lithology), geomorphological and seismic factors. The landslide is very close to the NAMA fault. Therefore, it is assumed to have suffered strong tremors during the previous earthquakes. In addition, the slope escarpment, high local relief, the approximation of the fault, active tectonics and densely articulated strata all favor the reproduction of landslides [21-23].

Although precipitation and climatic conditions can contribute to the triggering of landslides, we exclude this possibility because the study area is located in a Mediterranean climate zone, with an average annual rainfall of less than 600 mm. The paleoclimate records and modern meteorological observations show that the water vapors in the Middle Atlas and its adjacent areas mainly originated from the Mediterranean Sea [9]. In the upstream section of the Tamourghout River, the climate was relatively warm and humid. However, the temperature and precipitation showed noticeable decreases. The exposed rocks on both sides of the Maghraoua area are mainly hard volcanic rock intercalated with Limestone, which is well cemented and stable under natural conditions. During the late Holocene period, the study area was dominantly characterized by minimal rainfall and a dry climate [9]. Therefore, it could be inferred that it is less likely that any large dam-forming landslides (rock avalanches) be induced by rainfall.

Conclusion

Field investigation and geomorphic analysis were conducted to better understand the Tamda landslide formation process in Tamourghout river valley, Middle Atlas, Morocco. The integrated utilization of data obtained revealed that the large dams formed in the Tamda lake are housed in a picturesque site where a vast catastrophic landslide blocks the valley. It attests to the importance of abrupt natural events in the shaping of the relief. These events (collapse, landslide) would have been generated in Tamda elsewhere by the recent tectonic activity. Moreover, it turns out that the lithology and the flooded structure in the Middle Atlas are the main factors to explain the landslides genesis. Climatic conditions (humidity and cold) added to those of the vigorous tectonics were able to generate mass instabilities, capable of obstructing valleys as wide and as deep as that of the Atlas and subsequently formed the landslide-dammed lake.

References

1. B. Fedan, « Etude Structurale D'une Portion De L'accident Nord Moyen Atlasique Dans La Region Au S'v De Boulemane, Moyen Atlas Central, Maroc », p. **10**.
2. [G. Colo, Contribution a l'étude du Jurassique du Moyen Atlas septentrional: Atlas de planches hors texte. Éd. de la Division de la géologie, Direction, Ministère, Royaume du Maroc, 1961.
3. A. Akasbi, D. Sadki, A. Akhssas, et B. Fedan, « Dynamique sédimentaire et contrôle tectono-eustatique des dépôts dans l'intervalle Toarcien supérieur-Bajocien inférieur du Sud-Est du Moyen Atlas plissé (Maroc) », p. **8**.
4. H. Termier, « Etude géologique sur le Maroc central et le Moyen Atlas septentrional: Notes et Mémoirs du Service des Mines et de la Carte Géologique », Rabat Maroc, no 33, p. **4**, 1936.
5. M. L. Arboleya, A. Teixell, M. Charroud, et M. Julivert, « A structural transect through the High and Middle Atlas of Morocco », J. Afr. Earth Sci., vol. 39, no 3, p. **319-327**, juin 2004, doi: 10.1016/j.jafrearsci.2004.07.036.
6. R. du Dresnay, « Données stratigraphiques complémentaires sur le Jurassique moyen des synclinaux d'El Mers et de Skoura (Moyen-Atlas, Maroc) », Bull. Société Géologique Fr., vol. 7, no 6, p. **883-900**, 1963.
7. « Khalid OBDA et al..pdf ». Consulté le: mars 19, 2021. [En ligne]. Disponible sur: http://www.geo-maghreb.com/journal/volumes/Geomaghreb5/Khalid%20OBDA%20et%20al..pdf.
8. D. Sadkaoui et al., « Indices néotectoniques dans les sédiments du lac Ifrah (Moyen Atlas, Maroc) », Eur. Sci. J., vol. 11, no 14, 2015.
9. A. Taous, A. Tribak, K. Obda, R. Baena, E. López-Lara, et J. Bonilla, « Karst et resources en eau au Moyen Atlas nord-oriental », Geomaghreb, vol. 5, p. **41-59**, janv. 2009.
10. J. Martin, « Le Moyen Atlas central, étude géomorphologique », 1981.
11. S. Mounir, N. Saoud, M. Charroud, K. Mounir, et J. Choukrad, « The Middle Atlas Geological karsts forms: Towards Geosites characterization », Oil Gas Sci. Technol. D'IFP Energ. Nouv., vol. 74, p. **17**, 2019.
12. K. J. Hsu, « Catastrophic debris streams (sturzstroms) generated by rockfalls », Geol. Soc. Am. Bull., vol. 86, no 1, p. **129-140**, 1975.
13. A. Sabaoui, « Rôle des inversions dans l'évolution méso-cénozoïque du Moyen Atlas septentrional (Maroc). L'exemple de la transversale El Menzel-Ribat Al Khayr-Bou Iblane », Thèse Sci., **432**p, 1998.
14. T.-E. Cherkaoui et L. Asebriy, « Le risque sismique dans le Nord du Maroc », Trav Inst Sci, p. **225-232**, 2003.
15. T. E. Cherkaoui, A. El Hassani, et M. Azaoum, « Impacts du tremblement de terre de 1755 au Maroc: histoire, société et religion. In «Witnesses of Chaos: aspects of the 1755 Lisbon earthquake» », Acad. Ciênc. Lisb., p. **53-68**, 2017.
16. B. El Fellah, « Guelta Tamda (Moyen Atlas Plissé, Maroc): un exemple de lac de barrage naturel », Rev. Géographie Maroc, vol. 16, no 1-2, p. **115-126**, 1994.
17. B. El Fellah, O. Azzouz, et L. Asebriy, « Sikha Asfalou: exemple de glissement sur la côte méditerranéenne des Bokoya entre Torrès et Badis,(Rif septentrional, Maroc) », Bull.-Réseau Eros., no 16, p. **222-231**, 1996.
18. T.-E. Cherkaoui et L. Asebriy, « Le risque sismique dans le Nord du Maroc », Trav Inst Sci, p. **225-232**, 2003.
19. I. Etebaai, « L'environnement actuel et le fonctionnement hydroclimatique de quelques systèmes lacustres dans le Moyen Atlas marocain: cas des lacs Ifrah, Iffer et Afourgagh », Universite Abdelmalek Essaadi Faculté des Sciences et Techniques Tanger, 2009.
20. B. Damnati, I. Etebaai, H. Benjilani, K. El Khoudri, H. Reddad, et M. Taieb, « Sedimentology and geochemistry of lacustrine terraces of three Middle Atlas lakes: Paleohydrological changes for the last 2300 cal BP in Morocco (western Mediterranean region) », Quat. Int., vol. 404, p. **163-173**, 2016.
21. J. E. Costa et R. L. Schuster, « The formation and failure of natural dams », Geol. Soc. Am. Bull., vol. 100, no 7, p. **1054-1068**, 1988.
22. [G. B. Crosta, P. D. Negro, et P. Frattini, « Soil slips and debris flows on terraced slopes », Nat. Hazards Earth Syst. Sci., vol. 3, no 1/2, p. **31-42**, 2003.
23. D. J. Varnes, « Landslide Types and Processes », p. **28**.

Modeling the risks of marine submersion on the Saidia coastal-Cap de l'Eau (Northeast Morocco)

Samhane Ramdani[1], *Mostafa* Oujidi[1], *Najat* Amarjouf[2], *Hafssa* Afif[1], and *Abdellah* Azougay[1]

[1]Applied Geosciences Laboratory (LGA), Department of Geology, Faculty of Sciences Oujda - FSO, Mohammed Premier University.
[2]Research and Development in Engineering Sciences Laboratory "LRDSI", Department of Earth and Environmental Sciences, Faculty of Science and Technology Al Hoceima - FSTH, Abdelmalek Essaadi University, Tetouan.

Abstract. The coastline is an area of ecological and environmental balance between the continent and the sea. The coastline from Saidia to Cap de l'Eau, located on the Mediterranean coast of Morocco, is the subject of this study. The goal of this work is to produce maps of the risk of marine submersion. The methodology followed consists of entering storm meteorological data and land use data into the Iber 2D software. In addition, we mapped the hazard as well as the vulnerability of this submersion. The results of this 2D modeling of the hazard have shown that the coastline of Saidia - Cap de l'Eau presents a high risk of marine submersion, and this by the water levels rises to more than 4 meters and overflow on a width of more than 20 meters towards the mainland. In addition, the coastal strip constitutes a zone of high vulnerability compared to the continent.

1 Introduction

The Mediterranean coastline is exposed to several natural hazards that promote its vulnerability and degradation following the concentration of projects and socio-economic activities that involve human issues, property, and infrastructure, facing a significant influence to a risk of marine submersion, coastal erosion, and even a tsunami-genic risk. The coastline of Saidia - Cap de l'Eau extends over 20 km, which presents a pole of attraction and essential investment of national interest. [1]

Presumably, this coastline is threatened by a major risk of marine submersion that could invade a vast area involving human lives, property, and infrastructure. The existence of this risk of marine submersion, which consequently contributes to a specific rise in water level along the said coastline, is well illustrated in the 2D modeling of the hazard carried out on Iber software.

2 The study area

The study area extends from Saïdia to Cap de l'Eau, is located in the North-East of Morocco. It has a 20 km long coastline, limited by the Mediterranean Sea in the North, the Ouled Mansour Mountains and Triffa Plain in the South, the Kiss stream on the Moroccan-Algerian border in the East the Kebdana in the West (Fig. 1).

The Saidia-Cap de l'Eau coastline enjoys a rather contrasting Mediterranean climate, marked by an increasingly accentuated aridity, an average annual rainfall of 320mm/year, and an average annual temperature of 17°C.

The coastline of Saidia - Cap de l'Eau is generally constituted by the current to recent quaternary deposits and middle and ancient continental quaternary, plio-quaternary volcano-sedimentary, and the presence of a neogenous volcanism. [2]

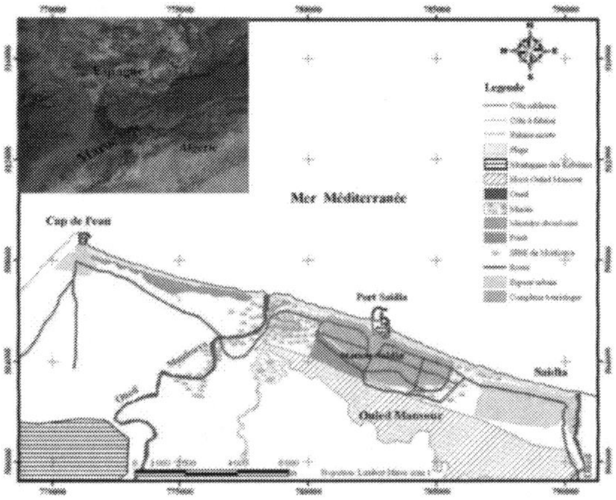

Fig. 1. The map of the situation of the study area [2]

This coastline is dominated by swells which are of North-West and North-East direction. Their interference generates a dominant littoral drift with a WNW-ESE direction. This is evidenced on the left bank of the mouth of the Moulouya by the coastal arrow very developed parallel to the shoreline characterizing the coastline dominated by the swell.

* Corresponding author: smahane.ramdani@ump.ac.ma

3 Materials & Methods

The present study concretizes the existence of the risk of marine submersion at the level of the littoral of Saidia - Cap de l'Eau, by making the conjunction of the hazard, the vulnerability, and the stakes, and this by pursuing an appropriate scientific protocol which is based on a treatment of the system of geographical information on ArcGis, and modeling in 2D on Iber software for the elaboration of the maps and the relevant results. [3]

Table 1. Primary data for the realization of the modeling

	Resolution/ coordinates	Softwares	Data used
DEM	12.5m	Global Mapper	Digital model and terrain topography
Satellite Imagery	35.09 -2.24	Mapuzzle	Land use and vulnerability
Storm Data	-	Excel workbook	Wave height as a function of time

The table above summarizes the primary data needed to model the risk of marine flooding on Iber 2D software and its characteristics.

The work methodology follows an explicit scientific protocol that addresses the path of the elaboration of the maps as a result. It is subdivided into two (02) parts; the first one on ArcGis contributes to preparing the data and their processing, and the other one on Iber realizes the 2D simulation of the marine submersion hazard. In ArcGis, the primary data processing is carried out, which is the object of the first step, and then the second part is started, whose main goal is to carry out the 2D simulation in Iber. [4] [5] and [6].

The indicated approach must be carefully carried out to achieve the main objective of this study, which is based on the simulation and the elaboration of different maps that define the marine submersion hazard, the risk intensity, and the stakes.

4 Results and Discussions

The results obtained are the crucial object of this study, divided into two essential parts: one carried out on ArcGIS and the other on Iber following an appropriate methodology. This study led to the elaboration of several maps, which highlighted the different entities that compose the area and its vulnerability.

Using ArcGIS software, we determined the land use map that summarizes the most important entities on the site: river, dense vegetation, forest, brushland, infrastructure, sand, and residential, as illustrated in figure 2.

This will subsequently help realize the vulnerability map, which is defined by three classes of vulnerability (0 - 0.5 - 1) classified by order of importance, as illustrated in figure 3.

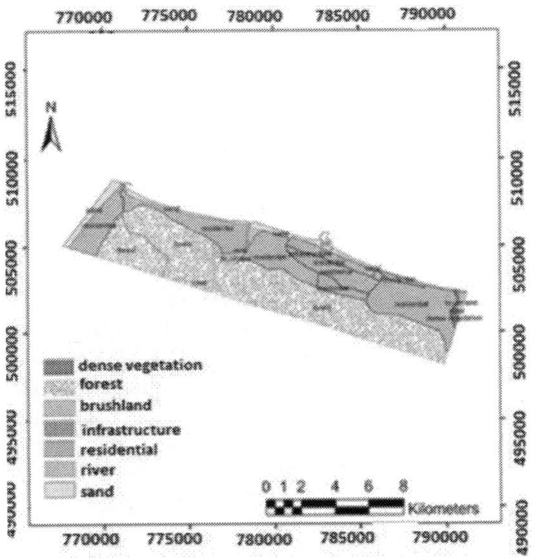

Fig. 2. The map of the land-use study area

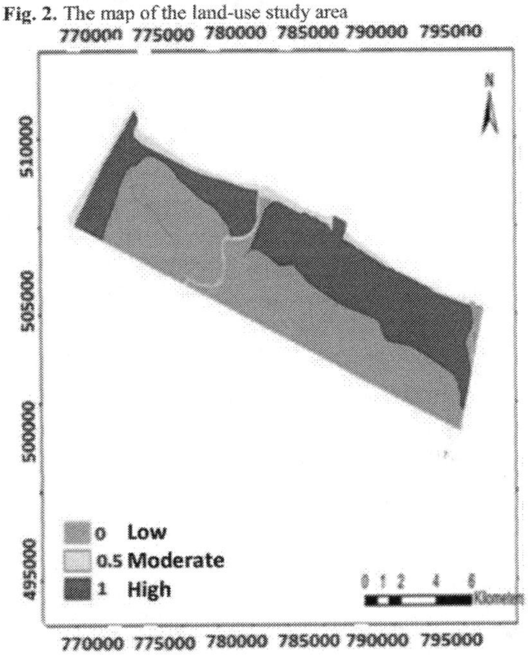

Fig. 3. Map of the vulnerability study area

In order to refine the results obtained, the study area is divided into two sectors: (figure 4)

Sector 1:

- From Saidia to the mouth of the Moulouya:

The 2D modeling carried out on Iber has allowed obtaining a result composed of three parameters: (1) water velocity, (2) water depth, and (3) water elevation, which predefine the intensity of the risk of marine submersion as well as the vulnerability of the concerned coastline. In addition, other factors condition the risk of marine submersion, namely: the topography of the land, the intensity of the topography of the land, the intensity of the meteorological storm that will favor the elevation of marine waters.

Sector 2:

- From the mouth of the Moulouya to Cap de l'Eau:

The results obtained show that we have a high vulnerability near the coastline from the mouth of the Moulouya arriving at Cap de l'Eau Moreover, or the depth of water varies between 2m to 0 m from the sea approaching the coastline and is well-illustrated in the area so close to the mouth of the Moulouya.

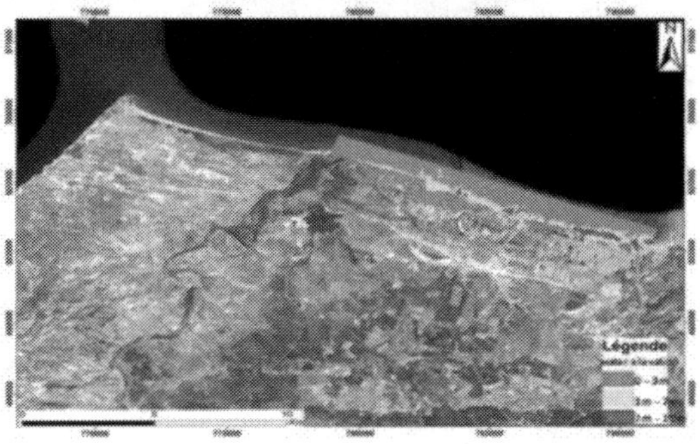

Fig. 4. The map of water elevation

Figure 4 illustrates the simulation result on Iber 2D, which shows a rise in water along the coastline following a storm that will promote the amplification of wave heights.

5 Conclusion

The risk of marine submersion would impact the coastline of Saidia-Cap de l'Eau and would have caused some damage that would generally affect the low points of the coastline. Hence the rise in sea level, in connection with global climate change. Thus, the rise in sea level will undoubtedly increase the recurrence of this risk. This work of mapping by two-dimensional hydraulic modeling has shown us good results regarding the risk of marine submersion on the eastern coast of the Mediterranean. Moreover, that consequently can generate an overflow of marine waters of a speed that can reach 3m / s during an assignment of the coastline by the hazard of marine flooding.

Acknowledgment

The authors would also like to extend their sincerest gratitude to the anonymous reviewers for their thoughtful comments and suggestions for improving this manuscript.

References

1. MOUZOURI M., IRZI Z., Évolution et morpho-dynamique de la plaine côtière de Saïdia (littoral méditerranéen du Nord-Est du Maroc durant la période 1958-2006). Bulletin de l'Institut Scientifique, Rabat, section Sciences de la Terre, 2011, N°33, p. 65-76.
2. LASGAA H; BOUMEAZA T; SBAI A; Le littoral méditerranéen du nord-est du Maroc entre dégradation du patrimoine naturel et nécessité de la Gestion Intégrée des Zones Côtières. Actes de la table ronde « La Gestion Intégrée des Zones Côtières », 2010. P. 40-54
3. A. M. Da COSTA LOMBA, OUJIDI M, AMARJOUF N, Apport de la modélisation 2D dans la cartographie préliminaire de l'aléa de submersion marine, cas de la côte de Luanda, Angola (Afrique Ouest), (3ème édition du colloque international des utilisateurs du SIG, P27. 2016.
4. AMARJOUF, N., "Interactions entre crues-ruissellements et aménagements de territoire et leurs impacts sur les versants Nord de la péninsule de Guelâaya: Cas de la baie de Melilla Béni Ansar et la baie de Betoya." (Maroc Nord Oriental).Thèse de Doctorat, Faculté des sciences, Université Mohamed Premier, Oujda. 302. 2018.
5. IBER, (a). Hydraulic Reference Manual for IBER V1.0. 2010
6. IBER, (b). Tutorial; Tossa. 2010.

Flood Hazard Mapping in Rural Area: Case of Ephemeral Watercourses in SidiBoulenouar Zone

Imane Amar[1,*], *Said* Bengamra[1], *Najat* Amarjouf[2], and *Abdellah* Azougay[1]

[1]Applied Geosciences Laboratory "LGA", Department of Geology, Faculty of Sciences Oujda - FSO, Mohammed Premier University, Oujda.
[2]Research and Development in Engineering Sciences Laboratory "LRDSI", Department of Earth and Environmental Sciences, Faculty of Science and Technology Al Hoceima - FSTH, Abdelmalek Essaadi University.

Abstract. The eastern region of Morocco is characterized by an arid to semi-arid climate. It is often confronted with heavy rainfall episodes that can generate floods that can cause adverse human and property damage. This work studies hydrological risk in the SidiBoulenouar area (Oujda province) to model floods and map risk areas. The method used is to determine the watersheds and their watercourses, then to calculate the flows with the mathematical formulas. Topographic data were extracted from a 12.5-m resolution DEM derived from the ASTER (Advanced Spaceborne Thermal Emission and Reflection Radiometer). GDEM (Global DEM). The DEM was used to delimit the hydrographic network with ArcMap software. The use of HEC RAS 1D was necessary to delimit the flooding area. The results of the 1D hydraulic modeling of the centennial flood showed that four ephemeral watercourses landlock the rural area of SidiBoulenouar at the time of the floods.

1 Introduction

Water is becoming increasingly important in our lives. However, it requires rational management to control its scarcity, but also its excess. This management is essential to ensure the supply of drinking and industrial water, agricultural needs, energy production, and the maintenance of the quality of our environment. This management requires the development of adequate methods and tools to minimize flood risks [1]. In terms of natural disasters worldwide, floods are the leading cause of natural disasters, claiming around 20,000 victims yearly [2]. The hydrological risks in Eastern Morocco are caused by severe floods resulting from heavy rainfall over a limited area in a short period. These precipitations are, in most cases, sudden and have a relatively high peak flow. [3].

The area of SidiBoulenouar situated in Oujda Province is drained by the two watercourses: Hassa creek and Abidine creek. These two creeks are temporary and generally flow during rainfall. They provoke floods at each thunderstorm, which disrupts the activities of the local population. The objectives of this work are flood modeling and spatial mapping of flooded areas in the center of SidiBoulenouar and its immediate neighbors by using the HEC RAS 1d model. This study was based on identifying the hydrological characteristics of the watersheds, calculating concentration times and peak flows for different return periods, the simulation of floods to determine the areas at high risk of flooding, and consequently, the proposal of floods civil engineering developments.

2 Geographical and Geological context of the Study Area

The centre of SidiBoulenouar is constituted by the douars of Oulad Ali Ben Yahya, Oulad Ben Gana, OuladBouhasaka and the seguia of the Guernaz stream which are part of the rural commune of SidiBoulenouar. This rural commune is bounded to the north and east by the rural commune of SidiMoussaLamhaya, west by the rural commune of Mestferki, and to the south by the province of Jerada. The commune is administratively attached to the Caïdat of Isly. The villages are located southwest of Oujda, about 20 km as the crow flies and accessible from Road N 17 and then by a track crossing the Isly stream. The study area is located in the Oujda-Taourirt corridor (eastern Morocco), bordered to the north and south, respectively, by the BeniSnassen belt and the Oujda mountains. This sector covers an area of about 900 km². From a geological point of view, the western mountain ranges of BeniSnassen and the mountains of Oujda are made up of a Paleozoic basement (schist and granite) surmounted by the Triassic volcano-sedimentary complex [4], which is covered by Jurassic carbonate formations. The Naima-El Aioun corridor is a plain corresponding to a large subsidence basin in which sediments of tertiary and quaternary age [5].

* Corresponding author: amar.imane@ump.ac.ma

3 Materials and Methods

We started with a hydrological study based on calculating the physical parameters (Perimeter, Area, Length, and Width of the Equivalent Rectangle, Length of the main watercourse, Gradient) of the different watersheds and watercourses crossing the study site. For this purpose, we processed the 30-meter resolution, digital terrain model using ArcGIS. Calculating the time of water concentrations in a watershed is defined as the maximum amount of time it takes for a drop of water to travel the hydrological path between a point in the watershed and the watershed outlet. The time of concentration is estimated using the most widespread formulas in Morocco (Formulas of Turazza,

Ventura, Kripich, D. Californian, H. Giandotti, Passini) [6] in addition to US Corps and Spanish. Then the statistical processing of rainfall data, peak flows for different return periods with different methods in the literature such as empirical formulas that consider two essential factors the size of the watershed and topographic Gradient.

The Gradex method is widely used for estimating project debit from rainfall data because, in most cases, the flow samples are much less extensive than rainfall samples. This means that it will be challenging to extrapolate return times of 10 or 20 years based on the flow sample. On the contrary, as rainfall samples are generally longer, it will be easier to determine the centennial value accurately. The fronco-Rodier method is an estimate of flows by analogy with other similar and adjacent watersheds. The criteria for similarity are essentially the morphology and rainfall regime.

The Hec-RAS established the flood simulation. Software developed at the Hydrologic Engineering Center of the US Army Corps of Engineers [7]. Calculations of water lines of gradually varying flows are based on Bernoulli's equation. The Manning-Strickler equation evaluates load losses for the friction term and by contraction-expansion coefficients. In abruptly varied flows, HEC-RAS uses the quantity of movement equation [7-8].

Finally, a hydrological study was carried out to determine flood debits using various empirical methods. These flows will be used in the hydraulic modeling carried out by the HEC-RAS software to simulate the flow of the studied rivers and delimit inundable zones.

4 Results and Discussion

4.1 Physical parameters of watersheds

The morphometric parameters of the different watersheds of SidiBoulenouar are summarized in Table 1. The four watersheds are high-altitude basins with limited surfaces and elongated shapes.

These characteristics were used in the calculation of concentration-time later.

Table 1. Physical characteristics of the four watersheds of SidiBoulenouar.

	Units	**Symbol**	**Basin Hassa**	**Basin Abidine**	**Basin Rivelet 1**	**Basin Rivelet 2**
Surface	Km²	S	2.20	1.41	0.14	0.05
Perimeter	Km	P	8.75	5.11	1.76	0.96
Compactness index		KG	1.65	1.20	1.32	1.22
Length of watercourse	Km	Leau	3.02	0.43	0.55	0.03
Equivalent rectangle length	Km	L	4.02	1.55	1.20	0.76
Equivalent rectangle width	Km	l	1.10	0.92	0.85	0.42
Maximum Altitude	M	Hmax	1195	960	796	772
Minimum altitude	M	Hmin	744	741	749	744
Height difference	M	Dth	451	219	47	28
Averageslope	%	Imoyenne	14.94	51.22	8.52	90.88

4.2 Estimation of concentration time Based

On the physical characteristics of the watersheds, we calculated the concentration-time by the different methods, which were then used to estimate the point flow (Table 2).

Table 2. The concentration time calculated by the various empirical methods in (h)

Formulas	Hassa Creek	Abidine Creek	Rivelet1	Rivelet2
Giandotti	0.62	0.46	0.42	0.22
Turrazza&Passini	0.53	0.13	0.16	0.01
Ventura	0.49	0.21	0.16	0.03
Kirpich	19.36	2.67	6.49	0.28
Espagnole	2.30	1.20	0.74	0.30
Van Te Chow	0.46	0.09	0.18	0.01
Californienne	0.71	0.10	0.24	0.01
US Corps	0.94	0.16	0.28	0.02
Chosen value	0.65	0.40	0.17	0.01

4.3 Simulation of flood flow in the area of SidiBoulenouar

Using HEC-RAS, the modeling of the flows generated bythe rivers gave the following results for the different return periods of 10, 20, 50, and 100 years (Fig. 1, 2, 3, and 4).

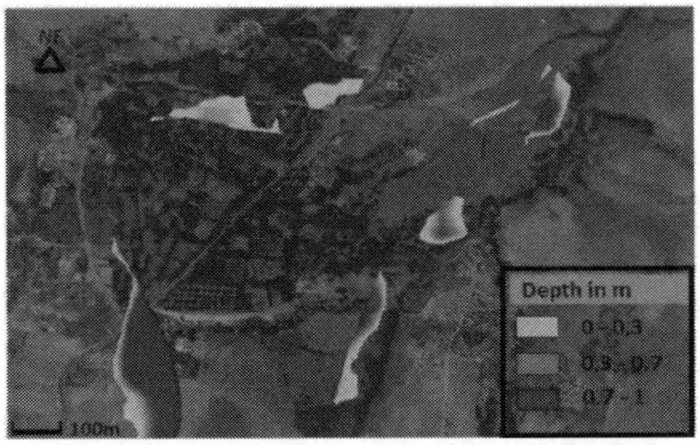

Fig. 1. Result of flood simulation in the study area for a 10-year return period

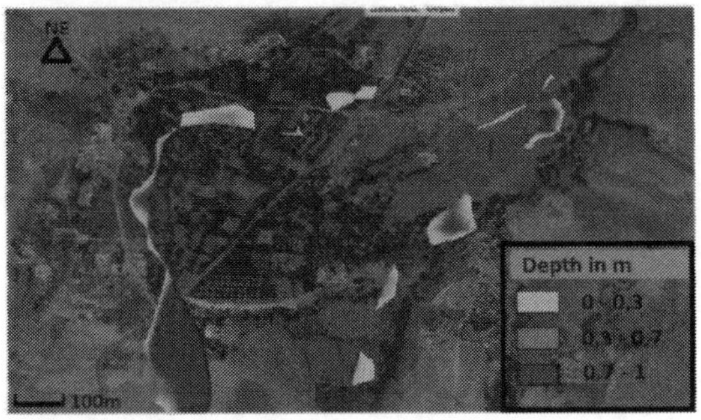

Fig. 2. Result of flood simulation in the study area for a 20-year return period

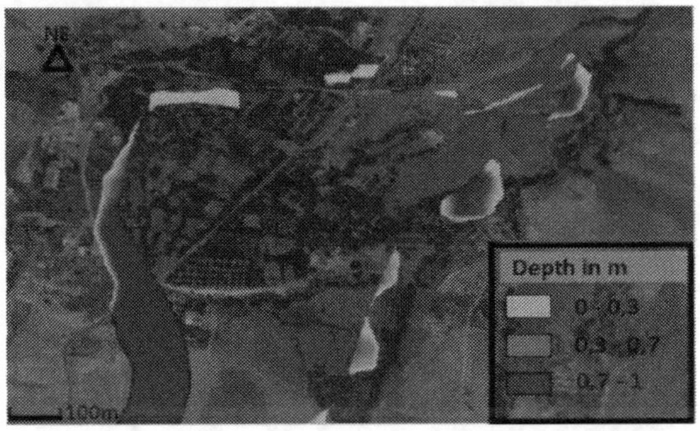

Fig. 3. Result of flood simulation in the study area for a 50-year return period

Fig.4. Result of flood simulation in the study area for a 100-year return period.

The overflows affect the intersection of the watercourses with the railroad tracks, farmland, and RP6030, which is not developed by any hydraulic structure. Flooding can also threaten the lives and property of the adjacent population. Therefore, three variants of hydraulic development have been proposed. (1) The diversion from Abidinecreek to Hassacreek upstream through a trapezoidal channel with a 5% slope and another channel downstream. (2) The second variant consists of sinking the buzzards on either side of the existing scuppers, and (3) the final solution is the construction of retention basins (Table 3).

- **Variant 1**

Table 3. The geometry of the trapezoidal channels according to the materials used.

		Canal 1	Canal 2
Channel 1	With the soil	1.5	2
	With concrete	2.5	1.5
Channel 2	With the soil	1.6	3
	With concrete	1.5	2.3

The slope of the rivers crossing the center of SidiBoulenoir is steep, averaging 5%, which will increase the speed of the flows and aggravate scouring, especially during the diversion of Abidinecreek towards Hassa creek. The insertion of para-digs to fight against scouring and, at the same time, waterproofing is necessaryin this case study.

- **Variant 2**

We kept the dimensions of the existing scupper without any variation, only the execution of the cleaning to emptyit, and we used the method of horizontal hydraulic jacking in order to insert the buzzard with a diameter of about 1000 mm so as not to modify the shape of the railway track, and we played only on their number to avoid any possible overflow

- **Variant 3**

A water retention basin is a storage area for rainwater, either underground or open-air. They are intended to temporarily store water to prevent flooding downstream in the watershed for the slow release of water.

A buffer zone can help manage fluctuations in rainwater input from these surfaces to release homogeneously into the environment.

5 Conclusion

Hydrological risk modeling in the SidiBoulenouar area is a methodical work that consists of following a well-defined methodology. It starts by identifying the characteristic information that emerges from the Hassacreek basin and the largest in terms of surface area 2.20km², followed by a hydrological study to describe the rain-flow transfer within the basins concerned and whose output

corresponds to the specific flows for each return period such as T=10years, T=20, T=50, T=100. These specific flows were used in a hydraulic study to describe the propagation of floods in SidiBoulenouar.

Acknowledgment

Special thanks go to Ryme EL AZZOUZ for his help. The authors would also like to extend their sincerest gratitude tothe anonymous reviewers for their thoughtful comments and suggestions for improving this manuscript.

References

1. S. Riad, 2003, "Typologie et analyse hydrologique des eaux superficielles à partir de quelques bassins versants représentatifs du Maroc," Thèse de Doctorat en cotutelle, Univ. Sciences et Techniques de Lille &Univ. IbnouZohr d'Agadir. **154** pages.
2. N. Simona Et L. Cedric, 2007. Gestion durable des zones inondables dans le Delta du Danube (Roumanie). 1res Journées Scientifiques Inter-Réseaux de l'AUF, Gestion Intégrée des Eaux et des Sols. Ressources, Aménagements et Risques en Milieux Ruraux et Urbains, Hanoi, Vietnam, **10** p.
3. N. Amarjouf, 2018. "Interactions entre crues-ruissellements et aménagements de territoire et leurs impacts sur les versants Nord de la péninsule de Guelâaya: Cas de la baie de Melilla Béni Ansar et la baie de Betoya." (Maroc Nord Oriental).Thèse de Doctorat, Faculté des sciences, Université Mohamed Premier, Oujda. **302** pages.
4. N. Rifai, A. Khattabi, And L. Rhazi, 2014, "Modelisation des crues des rivières pour la gestion intégrée du risque d'inondation: Cas du bassin versant de Tahaddart (Nord-ouest du Maroc)," Rev. des Sci. l'Eau. 27pages, pp. 57–69.
5. S. Bengamra And M. Oujidi, 2015 "Mineralogy of atmospheric dust deposits in the Naima-El Aiounbasin (Eastern Morocco)," pp. 2076–2082.
6. N. Amarjouf. M. Oujidi. D. Meijer. W. Klop. 2019. ISHHydrauliqueEngeneering journal. Evaluating ephemeral watercourse discharges by hydrologic-hydraulic modelling: application to the Kert River basin, Morocco. pp 1-11.
7. USACE., 2016. HEC-RAS Hydraulic Reference Manual. Version 5.0.3.
8. M. Lahsaini And H. Tabyaoui, 2018, "ModelisationHydraulique Mono Dimensionnel Par HEC RAS, Application Sur L'oued Aggay (Ville De Sefrou)," Eur. Sci. Journal, ESJ. 18, p. **110.**

E3S Web of Conferences **298**, 02001 (2021)
ICCR 2020

https://doi.org/10.1051/e3sconf/202129802001

Environmental management of coastal careers along the Doukkala-Abda linea

Toufik Remmal[1,*], *Halima* Jounaid[1], and *Fouad* Amraoui[1]

[1]GAIA laboratory, UHIIC, FSAC, BP 5366, Maarif, Casablanca, Maroc

Abstract. This study proposes an environmental management model for sand and gravel quarries along the coastline from Doukkala to Abda. It aims at the one hand; the creation of a regional database bringing together all the quarries to identify and characterize the materials exploited as well as the major impacts and risks associated with subsequent physiographic changes; and on the other hand to develop intervention mechanisms to prevent, reduce or even eliminate the harmful effects of abusive exploitation on the environment and public safety, while taking into account the specificities of the region.

1 Introduction

The opening and exploitation of quarries on a national scale are experiencing considerable growth due to the real estate and tourism boom and the policy of major structuring projects (ports, dams, highways, etc.). Nevertheless, several problems are raised regarding their uncontrolled management and exploitation, which can cause significant damage to the environment. The current issue in the sector is to reconcile the economic necessities of the supply of materials and the requirements of the environmental approach following the principle of sustainable development.

For most careers opened before 2002, no real data on the careers was exploited, and no legislation applied. Currently, environmental impact studies are a necessary condition for authorization procedures. However, they do not imply that the measures recommended in the study are effective and controlled. On the other hand, while awareness of the problems of career management has given rise to a new law on the exploitation of quarries, it lacks implementing texts to regulate the exploitation of quarries effectively.

In this perspective, the present study, devoted to the coastal quarries of the Doukkala-Abda coastal line, falls (Fig. 1). It is a question of arriving at a reasoned and controlled management of these quarries while maintaining a balance between socio-economic development and preserving the environment, namely protecting landscapes, sites, and sensitive natural environments.

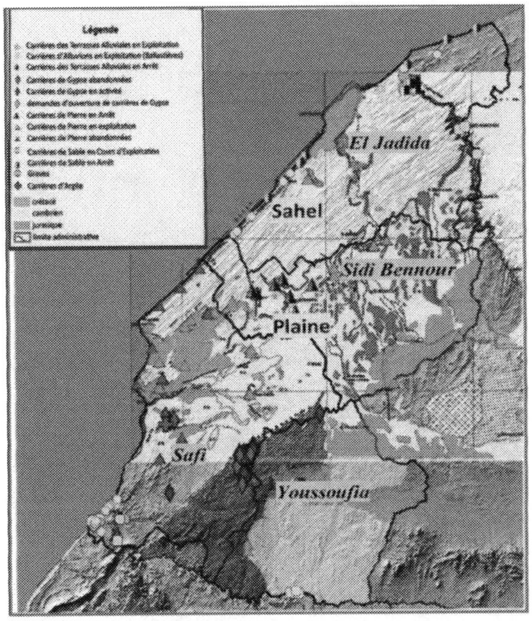

Fig. 1. The territorial distribution of inventoried quarries in the Doukkala-Abda provinces considers the region's geological and geomorphological structure.

2. Methodology

The approach undertaken consists of: i/ the constitution of a regional database necessary to identify and characterize the materials exploited, the quarry sites, the major impacts and the risks linked to physiographic modifications and ii/ the development of the mechanisms of intervention to prevent, reduce or even eliminate impacts on the environment and public safety while taking into account the specificities of the region.

*Corresponding author: remmaltoufik@gmail.com

https://doi.org/10.1051/e3sconf/202129802001

3. Inventory and land use

The region of Doukkala-Abda (DA) draws a set of approximately 7,700 km2, divided into two bands substantially parallel to the ocean [1]: The Sahel, endorheic coastal basin to the west, and the plain to the east (Fig. 1). The coastline stretches for about 350km, segmented into distinct morphological entities, including dune ridges locally flanked by marshes and mudflats, rocky cliffs with cliffs, beaches backed by stepped cliffs, and estuaries. The developments carried out on this shore respond, in general, to the socio-economic importance of the coastal zone in terms of urbanization, port infrastructure, processing industry, tourism, and fishing (Fig. 2).

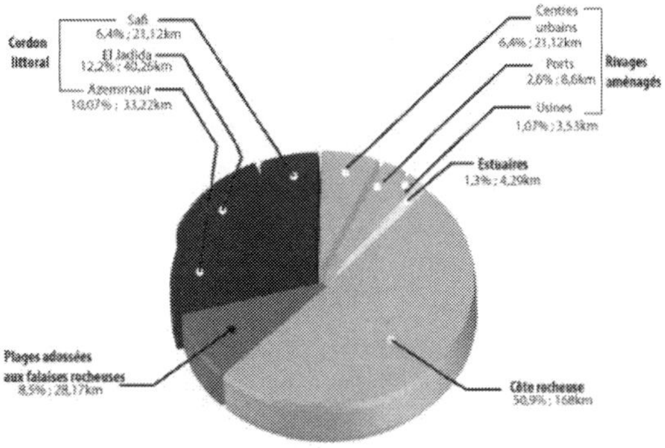

Fig. 2. According to the type of coast from Lamharza Essahel north of Azemmour to Lamaâchate south of Safi, distribution of the coastline is distributed according to the type of coast.

The histograms below (Fig. 3) show the distribution of quarries on the DA scale by category of materials, taking into account respectively the total volume authorized for mining and the overall surface area occupied by each type of quarry.

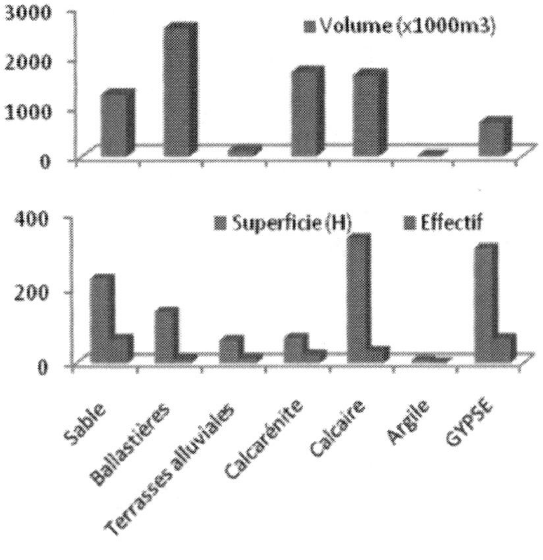

Fig. 3. Total area and volume authorized for extraction by category of material produced by the various quarries in the Doukkala-Abda region.

It emerges from this distribution that the land use by stone and sand quarries is higher correlatively to other materials. The limestone quarries show the greatest extension (~ 350 hectares) despite a small workforce, reflecting the extent of this material's quarry. In terms of volume, the ballast tanks provide the greatest production of substances useful for building and civil engineering work. They are the least numerous but settle on more than 120 hectares in the minor bed of the main river courses, namely Tensift and Oum Er Rbia. Sand exploitation exceeded 1.2 106 m^3, including more than 800,000 m^3 on the Sahel Doukkala coast alone. In this sector, extraction is carried out directly in the dune ridge, in the middle of the maritime public domain.

4. Major impacts due to the exploitation of quarries

The major impact problems posed by the various quarries in the region and the risks due to the physiographic modifications of the sites were assessed according to the three categories of materials, namely: the sand of the littoral rim, alluvial material on the river beds Oum Rbia and Tensift and the rocky stones of Sidi Bennour, Youssoufia and the province of Safi.

4.1 The sand quarries of the coastal barrier

The sand bars along the coast are a natural barrier against sea and wind erosion. Excessive exploitation of coastal sand irreparably leads to a modification of the hydrodynamic balance and the morphogenetic profile of the coast due to a deficit of alluvial material.

Three phenomena are consequent to this exploitation:
- The silting up of the land following the perforation of the coastal barrier and the disappearance of the plants that fix the sand at this level;
- Marine transgression, which leads to the replacement of the dune by seafronts;
- The washover phenomenon [2] is perceptible in the case of marshes and lagoons located behind the cordon, as in the classified Oualidia wetland. The increased erosion at the level of the lagoon causes a coastal drift, and the marsh risks being flooded by the burial of the lagoon silts under the deposits of marine sand, with the major consequence of the salinization of the water of the groundwater by mixing with the advancing seawater

4.2 Alluvial quarries

When the withdrawals from the stock of accumulated alluvial materials are greater than the inputs, the water line is lowered, which generates:
- Stream casing of the bed and destabilization of the banks by increasing their slope and undermining the base of the slope.
- Removal of engineering structures such as dams nearby.
- Decline of riparian vegetation by reduction of flood fields due to the lowering of the river bed.

- Modification or removal of habitat due to the lowering of the stream. The backwaters, the secondary branches, or the valley bottom meadows become flooded less frequently or for a shorter period and thus lose their interest in the proper functioning of the fluvial ecosystem.

4.2 Stone quarries

The major impacts noted are mainly visual and landscape, which results from the often artisanal mode of exploitation:
- Steep excavations and fronts near settlements (Sidi Bennour).
- Risks of the collapse of galleries (Safi clay quarries),
- Stacking of slag heaps (Safi gypsum quarries),
- Excessive occupation of space by crushing stations and the development of traffic lanes
Destruction and denaturation of the initial environment made up of arable or wooded land, especially the impacts of the non-rehabilitation of exploited sites.

5. Classification of operating sites

The environmental risks were assessed for each category of materials through criteria considering the severity of the environmental impact, the sensitivity of the receiving environment (natural parks, crop fields, protected areas, etc.), and compliance with the regulations and environmental strategies and programs.

The analysis made it possible to classify quarry exploitation sites according to three categories:

- Class 1: Areas prohibited for use

These areas benefit from strong legal protection (drinking water catchment area, biological reserve, wetlands, etc.) or present major negative impacts that require closure and possible rehabilitation.

- Class 2: Sensitive spaces.

These are areas of recognized fragility requiring monitoring during operation and rehabilitation in the long term. The authorizations to open and operate these quarries must necessarily be accompanied by specific requirements adapted to the degree of interest and the constraints.

- Class 3: Areas with minor issues

They do not require special provisions and regulatory provisions and can be opened for exploitation without great prejudice to the environment.

5. Regulatory aspects of quarrying

Until 2017, the exploitation of quarries was governed by centuries-old and obsolete regulations: the Dahir of June 1914, out of step with the new challenges and aspirations of the country. However, the legislator had initiated the first reform of this law in 2002, with Law N°. 08-01 relating to the exploitation of quarries, but which was not applied in the absence of the implementing texts. To remedy this situation, a circular from the Prime Minister was adopted in 2010, specifying, on a transitional basis, the procedures and modalities for the exploitation of quarries, pending the release of new regulations.

A new decree N ° 2-17-369 implementing the law N° 27-13 of 2015 relating to quarries was adopted, repealing the law of 1914. In detail, the draft decree includes provisions concerning the plans of quarry management, which will be broken down by region, with control of supply and demand and the cost of materials extracted from quarries. The definition of these quarry management plans will make it possible to locate the parts of the area where quarry exploitation cannot be authorized and to set the objectives to be achieved in terms of the redevelopment of quarry sites at the end of exploitation.

6. Career management

Despite the introduction of new regulations to put an end to a system perceived as unfair and favoring pensions and the informal sector, some aspects of fragility need to be removed:

The preeminence of the role of the supervisory administration, sometimes with a lack of clarity in interactions with the other departments concerned, reinforced by the absence of application texts and the multiplicity of references to other regulatory texts (regulations on the environment, regional planning, etc.);
- Lack of treatment of the problem of dune sands (looting of beaches), the public forest domain, or the beds of wadis.
- Lack of treatment of the financial (non-harmonized pricing policy) and tax aspects in the text;
- Non-availability of the implementing texts of the law on essential elements such as opening requirements and content of the specifications, composition, and mode of operation of the national commission and the provincial commissions for monitoring the management of careers, level of bank guarantee, follow-up register, etc.
- Need for the availability of human capacities and control tools, at the national and local level;
- Data relating to quarries are, for the most part, scattered among the various administrative services and insufficient to assess production by the quarry and by material, the rate of redevelopment or rehabilitation of closed or abandoned quarries, etc. It is necessary to know precisely the quantities exploited and the forecasts of material needs at the regional and national level to define the major orientations of the regional plan for the rational management of the materials exploited while respecting environmental constraints.

7 Rehabilitation of abandoned quarries

The rehabilitation of abandoned quarries involves rehabilitation operations and post-exploitation redevelopment, namely:
- Secure the site
- Correct disharmony in the landscape due to the location in the middle of built-up areas or the case of sprawl

- Provide a good environment by recreating a living environment adapted to the environment and consistent with the sector's development.
- Restore a vocation to the site (agricultural, forestry, industrial, tourist development)

While operations to restore and rehabilitate exploited quarries are the operator's responsibility, rehabilitation projects for abandoned quarries require consultation with all the partners concerned to adapt the sites to new uses by reconciling economic, environmental, and public safety aspects.

Thus, the identification of a quarry rehabilitation project requires a feasibility and impact study of the project, which must combine, on the one hand, a quality environmental approach and, on the other hand, meet the needs of economic development. Moreover, the social, of the municipality concerned and the wishes of the surrounding population.

Therefore, each rehabilitation project's financial, scientific, and technical implications must be carefully weighed to determine the long-term value of the project bearing in mind that in most cases, quarries are located in rural communities for which scarce resources. Human and financial resources would find better use in other development initiatives.

8 Conclusion

Environmental management of quarries requires site characterization, identifying sensitive areas. This rational operating plan respects the environment, a risk mitigation plan related to the operation, and a site redevelopment plan to rebalance the environment's ecological conditions. This management method is provided for in the text of the law, which also provides for control of the operation and rehabilitation of quarries.

Three types of activities can help reduce the impacts of exploiting AD quarries:
- The preservation of the most sensitive natural sites. (shorelines, river alluvium) by referring to the classification of sites according to environmental constraints, which imposes restrictions or even prohibitions for the most sensitive areas.
- The implementation of measures and processes to reduce the major impacts and reduce the nuisances generated by the activities. These measures concern the operation, rehabilitation or redevelopment, and the control and effective monitoring procedures to be implemented by operators and administrative authorities.
- A rehabilitation plan for old abandoned quarries to secure and/or give the site another vocation (re-greening, the transformation of excavations into lagoon basins, etc.)

Références

1. M. OUADIA : Les formations plio-quaternaires dans le domaine mésétien occidental entre Casablanca et Safi: Géomorphologie, Sédimentologie, Paléoenvironnements quaternaires et Evolution actuelle. Thèse d'Etat es Sciences (1998) . Univ. Mohammed V. Rabat, 319 p
2. A. Héquette et M.H. Ruz migration de cordons littoraux par processus de débordements de tempête dans le cadre d'une transgression marine, kvadehuk, péninsule de brôgger, spitsberg, Géographie physique et Quaternaire, 1986, vol. XL, n° 2, p. 197-206,

E3S Web of Conferences **298**, 02002 (2021) https://doi.org/10.1051/e3sconf/202129802002
ICCR 2020

Climate change and water resources management in the coastal sub-basin of El Jadida Safi, Morocco

Mounia El Azhari[1,*], and *Dalila* Loudyi[2]

[1]Hydraulic basin agency of Oum Er-Rbia (HBAOER), Morocco
[2]Univ. Hassan II of Casablanca, Faculty of Sciences and Techniques of Mohamedia, Morocco

Abstract. According to the 5th assessment report of the Intergovernmental Panel on Climate Change (IPCC), the temperature is projected to rise over the 21st century under all assessed emissions scenarios while precipitation will likely decrease, thus reducing renewable surface water and groundwater resources in most dry subtropical regions such as Morocco. Moreover, most projections of global circulation models predict a dryer future for North African countries. This work aims to assess the impact of climate variability on water resources in the coastal sub-basin of El Jadida-Safi, located in the Oum Er-Rbia basin. In order to determine future climate projections in the coastal sub-basin by the year 2099, we are using SDSM, a statistical downscaling tool based on both observed and anthropogenic emission scenario data for the pessimistic scenario RCP 8.5 and the optimistic scenario RCP 4.5. Results will help decision-makers and stakeholders better manage their water resources, prepare for extreme hydrological hazards, and enhance development planning in the river basin.

1 Introduction

Since 1950, changes in climate events have been observed worldwide, including an increase in temperature and a decrease in precipitations [1]. Climate change is projected to increase water demand, especially for irrigation in dry subtropical regions, and reduce surface and groundwater resources [2-7].

North African countries are expected to be the most affected by climate hazards, climate studies predict that temperature is to rise between 2 and 3°C while precipitation is to decrease between 10% and 20% until 2050 [8-9] (Paeth et al. 2009, Christoph et al. 2010, Schilling et al. 2012). Morocco already bears the effects of climate change, with a sequence of droughts or extreme precipitation events leading to floods. According to Modelling System for Agricultural Impacts of Climate Change (MOSAICC) which is a dynamic tool developed by the Food and Agriculture Organization (FAO) between 2013 and 2015, the average temperature could increase from 1.1 to 1.6 ° C by 2030, from 2.3 to 2.9 ° C in 2050, and from 3.2 to 4.1 ° C in 2080. At the same time, the precipitation could decrease by 14% in 2030, from 13 to 30% in 2050, and from 21 to 36% in 2080.

Water resources management studies require fine resolution climate data [10] (Timbal et al. 2009). However, General Circulation Models (GCM) have coarse resolutions, which make their application at the level of regional or local scales minimal. Consequently, downscaling methods were developed to span the spatial gap between the coarse GCMs outputs and the hydroclimatic basin variables [11-13].

In this study, future precipitation and temperature of the sub-basin were projected using a statistical downscaling approach under two RCP scenarios; results follow the trends of climate change projection worldwide. Some adaptation measures implemented at the level of the study area to face these imminent threats were discussed.

2 Study area

The coastal sub-basin of El Jadida - Safi is located southwest of the Oum-Er-Rbia basin. Its total area is of the order of 13070 km2. The basin includes the major cities of Safi, El Jadida, and Sidi Bennour, the industrial zone, and the port of Jorf Lasfar (Figure 1).

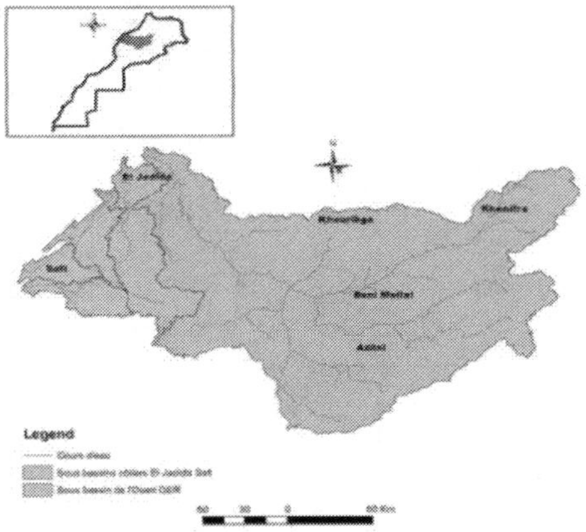

Fig 1. Location of El Jadida-Safi sub-basin

* Corresponding author: elazharimounia@gmail.com

A temperate coastal climate characterizes the sub-basin; annual precipitation varies between 200 and 300 mm, while temperature varies between 7.5 and 26°C. Surface water inputs to the study area are of the order of 50 Mm3/year. In addition, all non-consumed water from the upstream of Oum Er- Rbia basin estimated at 1590 Mm3/year are recovered by Al Massira dam and its compensators (Imfout, Daourat, and Sidi Said Maachou) and are transferred mainly to the coastal sub-basin of El Jadida – Safi. As to groundwater resources, they are generally contained in the aquifers of Sahel -Doukkala over an area of 6350 Km2 and allow the supply of more than 43 Mm3/year.

Mobilized water is mainly used to meet the needs for drinking and industrial water of the different urban and rural agglomerations in the sub-basin; it is also used to irrigate the Doukkala perimeter.

3 Statistical Downscaling

The statistical downscaling technique assumes that the relationships between the GCM outputs and the hydroclimatic catchment variables for the past observed climate are equally valid for the future, under changing climate [14]. In this study, multiple linear regression was used to project future temperature and precipitation at the level of the coastal sub-basin of El Jadida-Safi under two climate scenarios, RCP4.5, and RCP8.5.

The general equation for statistical downscaling based on the transfer function is:

$$R = F(X) \quad (1)$$

Where R represents the local climate variable, known as the predictand (precipitation, temperature), X is the set of GCM generated large-scale climate variables (predictors), and F is the stochastic or deterministic function that establishes the relationship between the predictands (R) and predictors (X) [15]. Using twenty-six predictors from the second generation Canadian Earth System Model (CanESM2) developed by the Canadian Centre for Climate Modelling and Analysis (CCCma) of Environment and Climate Change in Canada for RCP 4.5 and 8.5 scenarios for the period 1961-2100, we statistically downscaled both temperature and precipitation data taken from the meteorological station of Al Massira (Table 1). The predictors with the best correlation were chosen to project future climate from 2020 until the year 2100 (Table 1).

Table 1. Meteorological station characteristics

Station		Data period	Predictors	
Al Massira		1984-2017	Precipitation	Temperature
Longitude -7.63	Latitude 32.47		P8_fgl p8_ugl prcpgl	p500gl p850gl tempgl

Results of statistical downscaling for Al Massira station follow projections for the Oum Er-Rbia basin and Morocco in general with an increase in temperature and a decrease in precipitations. It will decrease by 3.21% for the RCP4.5 scenario and 3.44% for the RCP8.5 scenario for precipitation. In contrast, the temperature will increase by 2°C for the RCP8.5 scenario and by 2.2°C for the RCP4.5 scenario (Figures 2 and 3).

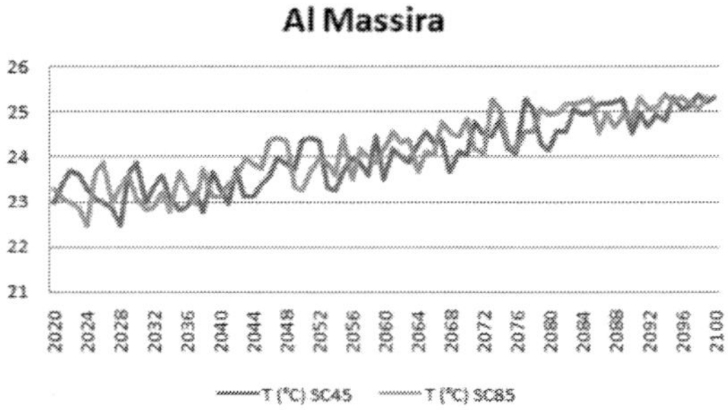

Fig 2. Evolution of temperature for both scenarios from 2020 to 2100

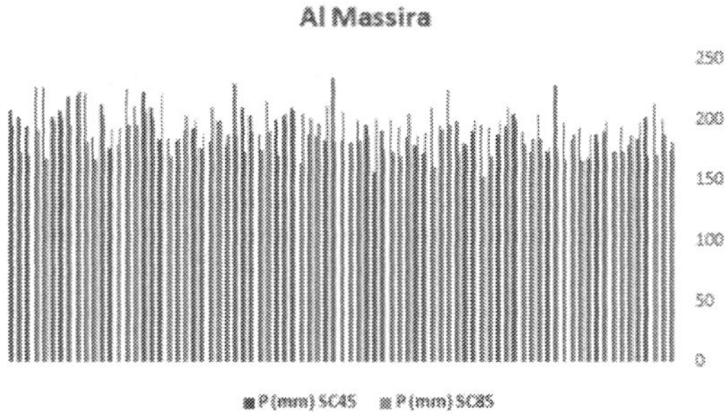

Fig 3. Evolution of precipitation for both scenarios from 2020 to 2100

Observed data over 37 years was used for calibration (1980-1998) and validation (1999-2017). The results of these two steps for precipitation and temperature are shown in Table 2. The values of R2 indicate a good correlation for temperature. However, for precipitation, and as already stated before, it is challenging to develop a perfect multiple regression equation because of the conditional behavior of precipitation.

Table 2. SDSM calibration and validation results for Al Massira station

Station	Predictand	R^2	
		Calibration	Validation
Al Massira	Precipitation	0.309	0.532
	Temperature	0.946	0.960

4 Adaptation measures

Considering the alarming climate projections in the sub-basin, and ensuring rational and sustainable management of water resources in the area to belay water availability for all users under acceptable conditions, policymakers and decision takers must integrate adaptation measures in programs and plans bound to water resources management. In recent years, at the level of Morocco and the study area in specific, several measures were implemented, such as the project of desalination of seawater launched by the Cherifien Office of Phosphates and which will reduce the excessive use of groundwater and allow the supply of 100 million m3/year through the plants of El Jadida and Safi. Besides, the Office of Agricultural Development of Doukkala adopted a strategy to rationalize and enhance the value of water use that involves several actions that limit water loss, improve performance of hydraulic structures and encourage users to save water by introducing new water-saving techniques leading to 40% of water savings [16-17].

5 Conclusion and recommendations

Results of statistical downscaling using SDSM for the sub-basin of El Jadida Safi confirm results and predictions of previous studies in the North African region. In fact, in the coming years, temperatures will tend to increase, while for precipitation, these will decrease with the alternation of dry and wet years.

These trends will indeed have a direct impact on water resources in the region. Thus, adaptation measures are necessary to maintain consistent offers to respond to different demands.

In order to quantify the impact of climate change on available water resources, the next step should focus on hydrological modelization of the sub-basin using available data and the results of this statistical downscaling.

References

1. IPCC, Climate Change 2014: Synthesis Report. Contribution of Working Groups I, II and III to the Fifth Assessment Report of the Intergovernmental Panel on Climate Change [Core Writing Team, R.K. Pachauri and L.A. Meyer (eds.)]. IPCC, Geneva, Switzerland, 151 pp (2014)
2. I.P. Holman. Hydrogeol. J. 14, 637–647 (2006)
3. B.E. Jiménez Cisneros, T. Oki, N.W. Arnell, G. Benito, J.G. Cogley, P. Döll, , T. Jiang, S.S. Mwakalila. Freshwater resources, 229–269 (2014)
4. X. Wang, J. Zhang, S. Shahid, E. Guan, Y. Wu, J. Gao, R. He. Mitig. Adapt. Strategies Glob. Change, 81–99 (2014)
5. K.E. Trenberth, J.T. Fasullo, TG, Shepherd. Nature Clim Change, 5, 725–730 (2015)
6. J.M. Kirby, M. Mainuddin, F. Mpelasoka, M.D. Ahmad, W. Palash, M.E. Quadir, S.M. Shah-Newaz, M.M. Hossain. Clim. Chang., 135, 481–491(2016)
7. J. Xia, Q.Y. Duan, Y. Luo, Z.H. Xie, Z.Y. Liu, X.G. Mo. Advances in Climate Change Research, 8(2), 63-67 (2017)
8. H. Paeth, K. Born, R. Girmes. Journal of Climate, 22, 114–132 (2009)
9. M. Christoph, A.H. Fink, H. Paeth. Climate scenarios. In: Speth, P., Christoph, M., Diekkrüger, B. (Eds.), Springer, Heidelberg, 402–425 (2010)
10. J. Schilling, K. P. Freier, E. Hertig, J. Scheffra. Ecosystems & Environment, 156, 12–26 (2012)
11. B. Timbal, E. Fernandez, Z. Li. Environmental Modelling & Software, 24(3), 341-358 (2009)
12. R.L. Wilby, T.M.L. Wigley. Prog Phys Geog , 21, 530-548 (1997)
13. R.L. Wilby, S.P. Charles, E. Zorita, B. Timbal, P. Whetton, L.O. Mearns. Supporting material to the IPCC, 3–21 (2004)
14. H.J. Fowler, R.L. Wilby, Water Resour. Res., 46, W03525 (2010)
15. H. Von Storch, E. Zorita, U. Cubasch. J. Climate, 6(6), 1161–1171 (1993)
16. S. Beecham, M. Rashid, & R.K. Chowdhury. Int. J. Climatol., 34, 3654-3670 (2014)
17. A. Guemimi, Séminaire sur la modernisation de l'agriculture irriguée, Rabat, Maroc. IAV Hassan II, 10 p. (2004)

E3S Web of Conferences **298**, 02003 (2021) https://doi.org/10.1051/e3sconf/202129802003
ICCR 2020

An Integrated Environmental Assessment of a coastal lagoon: The Case of the Oualidia lagoon (Morocco)

Mohammed Bouchkara[1,*], *Nezha* Mejjad[2], *Khalid* El Khalidi[1], *Donata* Melaku Canu[3], and *Bendahhou* Zourarah[1]

[1]Marine Geosciences and Soil Science Laboratory (URAC-45), Earth Sciences Department, Faculty of Sciences, Chouaib Doukkali University, BP 20, 24000, El Jadida, Morocco.
[2]PhD, associate member of Marine Geosciences and Soil Science Laboratory (URAC-45), Earth Sciences Department, Faculty of Sciences, Chouaib Doukkali University, BP 20, 24000, El Jadida, Morocco.
[3]National Institute of Oceanography and Applied Geophysics - O.G.S., Borgo Grotta Gigante, 42/c, 34010 Sgonico TS, Trieste, Italy.

Abstract. This paper aims to present a comprehensive socio-economic and environmental analysis of the Oualidia lagoon in Morocco. This lagoon is classified as a RAMSAR site since 2005, providing valuable ecosystem services for the surrounding population. This paper followed the Drivers-Activities-Pressures-State-Change-Impact-Responses (DPSIR) framework to evaluate environmental condition changes within the lagoon to support decision-making. The present research suggests a potential future model for safeguarding the lagoon's ecosystem without risking the well-being of future generations.

1 Introduction

Lagoons constitute 13% of the global coastline [1]. These are among the highest economically productive environments [2,3] and deliver numerous important environmental goods [4] that have a wide range of ecological, cultural, and socio-economic values which support several important natural services for human society [5,6]. They are widely regarded as the most precious components of the coastal zone [6]. However, coastal lagoons are undergoing substantial environmental degradation (Eutrophication, pollution, urbanization, and various forms of change) [7,8].

The Oualidia lagoon complex is one of the most important coastal wetlands in Morocco. It is essentially made up of a lagoon, swampy areas, and salt marshes, all occupying an interdunal furrow limited to the East by a cliff 50-80 m high and to the West by a cordon of living dunes. It is also home to various exploitations with great ecological value (artisanal fishing, aquaculture, exploitation of natural deposits of mollusks, sand production, etc.).

This lagoon is a site of biological and ecological interest and, since 2005, it has become an international Ramsar site. It provides valuable socio-economic and environmental benefits to its population. Thus, the Oualidia lagoon is one of the Moroccan ecosystems directly affected by human factors [9–12], which requires management interventions to achieve sustainable blue growth.

We present here the analysis of the results, including social, economic, and ecological aspects, through a holistic and integrated management framework (DPSIR).

2 Study Area

The Oualidia lagoon is situated on the Moroccan Atlantic coast. It is located 75 km south of El Jadida city (32° 44',42 N - 9° 02',50 W) (Fig.1). It is characterized by a warm temperate climate and an average rainfall, varying according to natural fluctuations in the rainfall regime.

This lagoon is among the most important coastal wetlands in Morocco. Today, in relation to its ornithological importance, this site has been classified as a site of biological and ecological interest (SIBE), through the Master Plan of Protected Areas and as a RAMSAR site since 2005. Of international value constitutes a migratory stopover and a winter refuge appreciated by the various species of water birds.

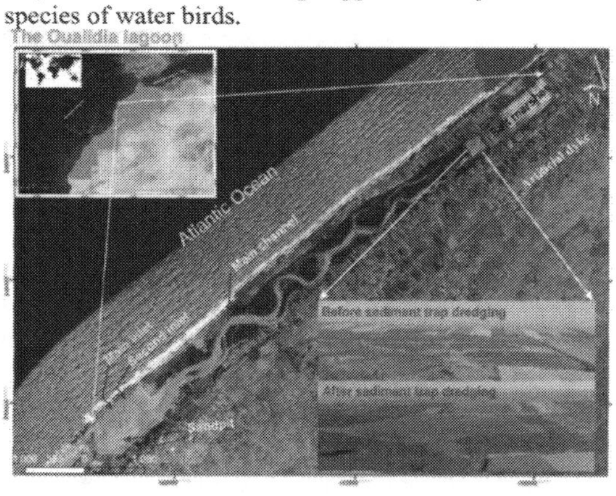

Fig. 1. The geographic location of the Study area

* Corresponding author: m.bouchkara@yahoo.com

3 Materials and methods

The socio-environmental analysis of the Oualidia Lagoon was evaluated based on the DPSIR framework as an integrative management process tool [13]. This framework originated in the late 1990s [14]. The Pressure-State-Response (PSR) framework was the early implementation of the DPSIR approach, which was suggested by [15] and further refined by the Organisation for Economic Co-operation and Development for Environmental Policy and Reporting [16]. Drivers-Pressures-State-Impact-Response (DPSIR) is a global issue structure model that the European Environment Agency has adopted in the 1999s to estimate the link between drivers, impacts, and policy responses [13,17]. The application of this model in our study is primarily based on a causal link between "driving forces" (economic sectors, population growth/land-use change), "pressures" (pollution, human activities), "states" (physical, chemical, and biological), and "impacts" on ecosystems, health, and human functions, ultimately leading to policy "responses" (managed projects, proposed projects).

4 Results and discussions

4.1 Drivers, Pressures, and Impacts on the Oualidia lagoon

The main anthropogenic drivers are the human activities practiced in and around the lagoon, as illustrated in (Fig. 2). Fishing, oyster culture, agriculture, urban and tourism activities are exercising many pressures on the lagoon, threatening its viability.

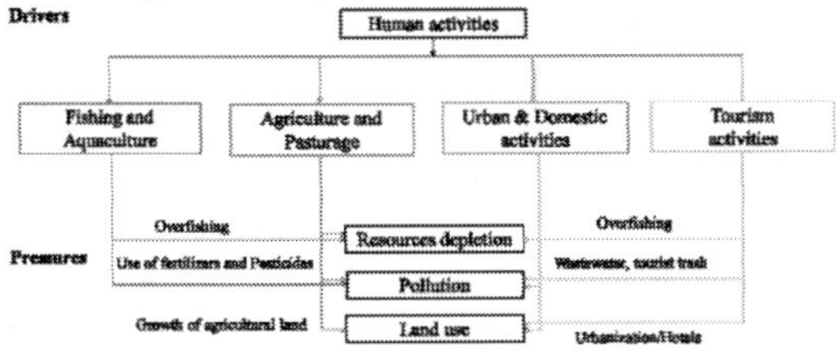

Fig.2. the main anthropogenic drivers and pressures in the Oualidia lagoon.

4.1.1 Aquaculture

The Oualidia lagoon is the historically important Moroccan aquaculture site. This activity is very relevant, especially oyster farming operated since 1970 in the lagoon [11]. Today, there are 8 oyster farms, and its annual production is about 200 tons of oysters. Nowadays, aquaculture activities encounter important issues caused by the deterioration of the environment due to various anthropogenic developments (organic and chemical pollution). This site, classified as Zone A, is now in Zone B and is at risk of being reclassified as Zone C.

4.1.2 Urbanization

Oualidia Lagoon has experienced an increase in land use and population concentration since 1950. The population density in Oualidia over the past 50 years has rapidly increased from 7,741 in 1971 to 13,076 in 1994, 15,430 in 2004, and 18,616 in 2014 (http://www.statistic.gov.ma/).

The change in land use can significantly affect runoff volume and heavy metal input (Maanan et al., 2014). Quantitatively, these effects are revealed for the entire watershed, and an assessment of land-use change was conducted by [11], who reported that in 2006, the main land cover types identified in the Oualidia Lagoon were agricultural areas (40.15%), salt marshes (25.29%), grasslands (20.45%) and urban areas (11.74%). The majority of the extension of urban areas was used for housing, businesses, institutions, infrastructure, and tourist areas [11].

4.1.3 Tourism and recreation

The Oualidia lagoon offers many tourist activities, including sailing, swimming on the beaches, bird watching, and nature observation. The seaside resort has a diversity of landscapes, and in summer, the population increases by 54,000 with a daily capacity of 30,000 people [18]. Touristic and recreational activities can cause environmental disturbances that can threaten biodiversity and ecological integrity.

4.1.4 Salt extraction

The salt marshes, covering an area of over 270 ha, play an important economic role. It is a process of artificialization of the wetlands of the lagoon, which started in 1949. The evolution of the salt marshes leads to the loss of wetlands containing the optimal conditions for biological life and the waste products abandoned there to distort the environment.

4.1.5 Agriculture

Farming is practiced in very narrow plots, on a strip bordering the wetland, both in the depressions and on the coastal dunes. In search of higher production, farmers remove fallow land from the agricultural rotation and intensify cultivation, using large quantities of chemical and organic fertilizers and phytosanitary products.

In the absence of a framework for rational input use, farmers overuse these products without respecting either the doses or the treatment periods. Drained by runoff water, all these products will contaminate the wetlands.

https://doi.org/10.1051/e3sconf/202129802003

Fig. 3. The main human activities in the Oualidia lagoon

4.2 Responses of Human society

4.2.1 Past Management projects

Despite the qualities mentioned above of the Oualidia lagoon, local populations growth and their activities intensification have been accompanied by a substantial degradation of their activities in the absence of safeguard plans values. Biodiversity is under continuous stress, as well as economic resources. However, the impacts are not limited to the exploitation of the lagoon complex; they come largely from degradation processes that extend over the watershed. Scientists believe that the severity of these impacts must be strongly emphasized, mainly because of the high sensitivity of the lagoon environments.

Morocco launched the HALIEUTIS plan in 2009. This new strategy for the development and competitiveness of the marine fisheries sector aims to ensure the sustainability of the sector through the preservation of fisheries resources and to meet the challenges of globalization by meeting increasingly demanding standards.

Similarly, Morocco created the National Agency for the Development of Aquaculture in 2011, which developed environmental measures followed by the aquaculture industries [18]. The National Agency for Aquaculture Development is the main actor in the aquaculture industry to promote sustainable aquaculture along the Moroccan coast, including in lagoons.

In 2011, Planète Urgence initiated a mission in Morocco to protect the lagoon of Oualidia as a RAMSAR site. This mission is conducted in partnership with two local associations, the Association for the Protection of the Oualidia Lagoon (APLO) and the Integrated Management of Resources (AGIR) located in Al Hoceima, Morocco. Several actions have been planned to consider all the environmental problems of the lagoon, namely intensive agriculture, excessive use of fertilizers and pesticides, irrigation, overexploitation of aquaculture resources, etc. These actions include an annual survey of migratory and endemic birds in the lagoon, the collection of macro-waste in the lagoon and on the surrounding beaches (this collection will be done by snorkeling), and participation in the inventory pelagic marine species around the lagoon of Oualidia (http://ma.chm-cbd.net/news/protection-de-la-lagune-d-oualidia).

4.2.2 Creation of a sediment trap

Considering many environmental problems such as the asymmetric nature of tidal propagation, the confinement of upstream waters, and the decrease in-depth, it was therefore advised in 2006 to create a sediment trap near the upstream dike in order to trap this sludge carried upstream by the strongest currents during the flow [19–22]. This sediment trap was constructed in 2011. Thus, the sediment trap construction facilitated the lagoon's hydrodynamics, especially in the sandpit area, which changed its morphology. This hydrodynamic resulted in a decrease in sedimentary organic matter upstream that is likely due to the trapping of fine sediments in the sediment trap [23].

4.2.3 Best practice recommendations for integrated management

Ecological monitoring system: Due to the importance of ecological conservation and its management, we propose to develop a simple ecological monitoring system as one of the main long-term measures, which will serve as a management and decision support tool. Thus, the installation of ecological monitoring stations is strongly recommended. These monitoring stations will be installed at different locations in the lagoon in order to continuously record all changes in the ecological parameters of the site (temperature, salinity changes, and other physicochemical factors), which could be the consequence of many human activities practiced at the site or of external natural trends such as climate change and morphological modifications of the inlets.

Monitoring of human activities: The economically viable natural resources of the lagoon of Oualidia are primarily derived from agriculture, aquaculture, fishing, tourism, and other associated activities.

- Fishing. Fishers use small-mesh boats and nets that allow them to catch fish of all sizes and shellfish. Therefore, the main pressures resulting from fishing are related to increasing the harvest of target fish and shellfish populations. This leads to their decline, which can also be exacerbated by illegal fishing and the use of illegal gear, lack of regulation, and insufficient controls. For this reason, the fishing activity must be monitored to highlight its different impacts on fish populations, birds, or the wetland environment in general.
- Agriculture. The monitoring system of this activity must allow the detection and evaluation of any impact on habitats (soil, water, plant groups, etc.) and species of fauna and flora.
- Tourism. Monitoring this economic activity will provide managers with information to justify these "uses" of space, which are not incompatible with nature conservation. In addition, EL Mahrad et al., 2020 proposed practical and feasible management measures to mitigate waste, including well-located trash cans, environmental advice for tourists, waste collection, and beach cleaning at the Oualidia lagoon in summer.

It is important to establish policy actions aimed at achieving a safe and unpolluted ecosystem, with a clear vision and feasible solutions, to present practical responses that contribute to reducing the degradation of the ecosystem's biodiversity and meet the needs of the local population.

5 Conclusion

The Oualidia lagoon is providing precious socio-economic benefits and ecosystem services for the local population. Nevertheless, the non-sustainable use of this ecological space has led to different types of degradation and numerous environmental problems.

The application of the DPSIR framework constituted a multidisciplinary approach that allowed us to outline the links between socio-economic activities, their impacts on the sustainability of the lagoon, and human interventions to reduce them, with a more strategic vision to improve the ecological productivity of the Oualidia lagoon, ensure the sustainability of its natural resources and integrated management of this ecosystem.

Responses were proposed as possible measures that could contribute to the sustainability of the Oualidia lagoon and its resources.

References

1. R. S. K. Barnes, Biol. Mar. Mediterr. **2**, 3 (1995)
2. R. W. Duck and J. F. da Silva, Estuar. Coast. Shelf Sci. **110**, 2 (2012)
3. B. Knoppers, in *Elsevier Oceanogr. Ser.* (Elsevier, 1994), pp. 243–286
4. M. Cañedo-Argüelles, B. J. Kefford, C. Piscart, N. Prat, R. B. Schäfer, and C. J. Schulz, Environ. Pollut. **173**, 157 (2013)
5. A. Newton, A. C. Brito, J. D. Icely, V. Derolez, I. Clara, S. Angus, G. Schernewski, M. Inácio, A. I. Lillebø, A. I. Sousa, B. Béjaoui, C. Solidoro, M. Tosic, M. Cañedo-Argüelles, M. Yamamuro, S. Reizopoulou, H. C. Tseng, D. Canu, L. Roselli, M. Maanan, S. Cristina, A. C. Ruiz-Fernández, R. F. d. Lima, B. Kjerfve, N. Rubio-Cisneros, A. Pérez-Ruzafa, C. Marcos, R. Pastres, F. Pranovi, M. Snoussi, J. Turpie, Y. Tuchkovenko, B. Dyack, J. Brookes, R. Povilanskas, and V. Khokhlov, J. Nat. Conserv. **44**, 50 (2018)
6. I. E. Gonenc and J. P. Wolflin, *Coastal Lagoons: Ecosystem Processes and Modeling for Sustainable Use and Development* (CRC Press, 2004)
7. E. Wolanski, A. Newton, N. Rabalais, and C. Legrand, (2013)
8. F. de A. Esteves, A. Caliman, J. M. Santangelo, R. D. Guariento, V. F. Farjalla, and R. L. Bozelli, Brazilian J. Biol. **68**, 967 (2008)
9. N. Mejjad, A. Laissaoui, O. El-Hammoumi, M. Benmansour, S. Benbrahim, H. Bounouira, A. Benkdad, F. Z. Bouthir, A. Fekri, and M. Bounakhla, J. Radioanal. Nucl. Chem. **309**, 1133 (2016)
10. N. Mejjad, A. Laissaoui, A. Fekri, A. Benmhammed, O. El Hammoumi, and E. K. Cherif, ACM Int. Conf. Proceeding Ser. (2020)
11. M. Maanan, A. C. Ruiz-Fernández, M. Maanan, P. Fattal, B. Zourarah, and M. Sahabi, Int. J. Sediment Res. **29**, 1 (2014)
12. B. Zourarah, M. Maanan, C. Carruesco, A. Aajjane, K. Mehdi, and M. Conceição Freitas, Estuar. Coast. Shelf Sci. **72**, 359 (2007)
13. M. Elliott, D. Burdon, J. P. Atkins, A. Borja, R. Cormier, V. N. de Jonge, and R. K. Turner, Mar. Pollut. Bull. **118**, 27 (2017)
14. E. R. Carr, P. M. Wingard, S. C. Yorty, M. C. Thompson, N. K. Jensen, and J. Roberson, Int. J. Sustain. Dev. World Ecol. **14**, 543 (2007)
15. D. J. Rapport, Stat. Canada 11-510, Ottawa, 1979 (1979)
16. OECD., O. de coopération et de développement économiques, O. Staff, and D. (OECD) Staff, *Environmental Indicators: OECD Core Set* (OECD, 1994)
17. M. N. Miranda, A. M. T. Silva, and M. F. R. Pereira, Sci. Total Environ. **718**, 134968 (2020)
18. B. El Mahrad, S. Abalansa, A. Newton, J. D. Icely, M. Snoussi, and I. Kacimi, Front. Environ. Sci. **8**, (2020)
19. H. Karim, E. Omar, I. Mohammed, L. Jamila, A.

Benyounes, E. Zineb, etOrbi Abdellatif., and M. Ahmed, Int. J. Adv. Res. **5**, 2015 (2017)

20. K. Hilmi, V. G. Koutitonsky, A. Orbi, J. I. Lakhdar, and M. Chagdali, African J. Aquat. Sci. **30**, 1 (2005)
21. E. T. Du, (n.d.)
22. K. El Khalidi, B. ZOURARAH, and A. AAJJANE, Estud. Do Quaternário **7**, 73 (2011)
23. A. Makaoui, M. Idrissi, A. Agouzouk, J. Larissi, T. Baibai, Z. El Ouehabi, M. A. Laamel, I. Bessa, O. Ettahiri, and K. Hilmi, Eur. Sci. Journal, ESJ **14**, 93 (2018)

https://doi.org/10.1051/e3sconf/202129803001

Assessment of the Coastal Marchica Watershed through Ecotoxicological Indices of Trace Elements in Superficial Sediments

Bouchra Oujidi [1,*], *Mohamed* Kabriti [2], *Mounia* Tahri [3], *Abdelfettah* Bensaber [2], *Abdelfettah* Benchrif [3], *Mostafa* Layachi[4], *Mohamed* Maanan[5], *Hocein* Bazairi [1], *Nadia* Mhammdi [6], and *Maria* Snoussi [1]

[1]Mohammed V University in Rabat, Faculty of Sciences. 4 Avenue Ibn Battouta, B.P. 1014 RP 10000, Morocco.
[2]National Laboratory for Studies and Pollution Monitoring (LNESP), Av. Mohamed Ben Abdellah Er-regragui Madinat Al-Irfane Rabat, Morocco.
[3]National Center for Energy, Sciences and Nuclear Techniques (CNESTEN), BP 1382, RP 10001 Rabat, Morocco.
[4]National Institute of Fisheries Research (INRH), 13Bd Zerktouni, BP 493, Nador, Morocco.
[5]University of Nantes LETG, UMR 6554, BP 81227, France.
[6]Mohammed V University in Rabat, Institut Scientifique, GEOPAC Research Center, Geophysic and Natural Hazards Laboratory, Morocco.

Abstract. This study analyses the seasonal distribution in the superficial sediments of the Marchica watershed of trace and major elements, grain size, and total organic carbon. Samples were taken during the wet and dry seasons of 2018 in the mainstream valleys carrying the discharges to the Marchica Lagoon. Ecotoxicological indices (Enrichment Factor, Contamination Factor, Pollution-Load Index) and sediment-quality guidelines were used to evaluate the ecotoxicological risk. The results revealed during both wet and dry seasons contamination by lead in all stream valleys. The northwest stream valley, characterized by some industrial activities, and the stream valley crossing the abandoned mine were the most polluted by zinc and copper. Minimal pollution for Cr, Co, and Ba was found during both the wet and dry seasons. Despite the restoration actions achieved around the Ramsar site of Marchica lagoon during the last decades, the watershed continues to have a negative impact on this ecosystem. Therefore, an environmental management strategy is necessary, taking into account the rehabilitation of the abandoned mining sites, the rational use of pesticides and fertilizers in agriculture, and the environmental responsibility of industrial companies, especially those installed before the implementation of the national law on environmental impact studies.

1 Introduction

Watersheds surrounding the coastal lagoons have experienced intense urbanization in recent years due to the ecosystem services offered by these ecosystems. However, this urbanization accompanied by the intensification of human activities around coastal lagoons has impacted their structure and function and compromised their ecological integrity [1-2]. Sediment flows in rivers reflect natural or human-induced changes in their watersheds [3]. The characterization of terrestrial sediment inputs to the seas is essential for watershed managers because of the harmful effects of sediments on nearshore marine ecosystems [4].

The Marchica watershed constitutes an example with a population of approximately 400 000 from cities around the Ramsar site of the Marchica lagoon. The watershed covers an area of 700 km^2 and is formed by four morphological structures, the volcanic massif of Gourougou in the northwest, the massif of Beni Boulffrour in the west; the Kebdana massif in the south; and the Bouarg plain in the southwest [5, 6].

The hydrographic network of the Marchica watershed is composed of several stream valleys called wadis, the majority of which have important contributions during rainy periods [7]. The geochemical assessment of the sediment inputs from the Marchica watershed, which are the main sources of pollution in the Marchica Lagoon, is crucial for protecting this Ramsar site and conserving its biodiversity.

2 Objectives

The objectives of this study are to assess Marchica watershed sediment characteristics through:

1. analysis of the seasonal distribution of trace elements (Pb, Cu, Zn, Cr, Co, Ba), major elements, grain size, and total organic carbon in the superficial sediments of the Marchica watershed;
2. ecological risk assessment of trace elements in superficial sediments;
3. identification of highly polluted areas.

* Corresponding author: bouchra_oujidi@um5.ac.ma

3 Methodology

Superficial sediments (0.5 cm) were sampled during the wet (March) and dry (July) seasons of 2018 in the main wadis of the Marchica watershed carrying the discharges to the Marchica lagoon through a representative station for each one (Fig. 1). The climate is Mediterranean with hot and dry summers and mild and rainy winters.

Major elements analysis was carried out using Bruker S1 Turbo SD hand-held X-ray fluorescence (HHXRF) spectrometer. Trace elements analysis were realized by an Agilent 4200 microwave plasma atomic emission spectrometer (MP-AES). Granulometry analysis was measured by a laser particle size analyzer (Malvern Mastersizer 2000) and Total organic carbon analysis by a LECO carbon analyzer. Analysis details are given in [8].

Enrichment Factor (EF) [9], Contamination Factor (CF) [10] and Pollution-Load Index (PLI) [11] were studied for trace elements (TEs) to access ecological risk:

EF=(CTEsl/CFe)Sample /(CTEs/CFe) Background (1)

CF=CTEs Samplen /CTEs Background (2)

PLI=(CF1 × CF2 × CF3 ×... × CFn)1/n (3)

Normalization using iron was calculated to quantify the anthropogenic metal pollution from the natural variability of trace elements, and reference values were taken from Rudnick and Gao (2003) [12]. Sediment concentrations were compared with sediment-quality guidelines (SQGs) for freshwater ecosystems to evaluate the ecotoxicological risk as reported by Macdonald et al. (2000) [13].

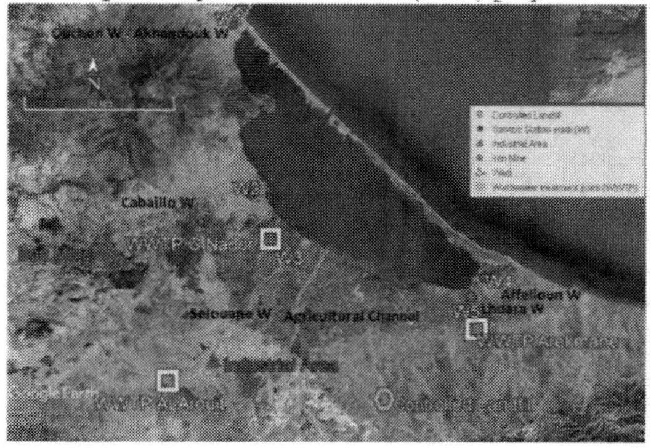

Fig. 1. Map of Coastal Marchica Watershed.

4 Results

4.1 Granulometry and TOC

The granulometric distribution sediment in the Marchica watershed showed that all sampling stations were classified as muddy sand facies during the wet season and sandy mud during the dry season except the first station, which was classified as sand during the wet season.

The TOC sediment results ranged from 0.70% to 4.58% during the wet season and from 1.82% to 10.10% during the dry season.

4.2 Major and Trace Elements

The average concentrations of major elements (%) during the wet and dry seasons showed the following pattern Si (17.63) > Ca (13.64) > Al (4.54) > Fe (3.90) > K (1.88) > Ti (0.39) > P (0.16) > Mn (0.04).

The average concentrations of trace elements (mg/kg) during the wet and dry seasons showed the following pattern Ba (450.81) > Zn (149.66) > Pb (144.9116) > Cr (50.04) > Cu (36.47) > Co (9.78).

5 Discussion

The ecotoxicological risk assessment used with SQGs and assessing the indices (EF, CF, and PLI) (Fig. 2, 3, and 4) revealed contamination by lead in all streams during wet and dry seasons valleys. The northwest stream valley, characterized by some industrial activities, and the stream valley crossing the abandoned mine sites were the most polluted by zinc and copper. Cr, Co, and Ba were found to be at minimal pollution during both the wet and dry seasons.

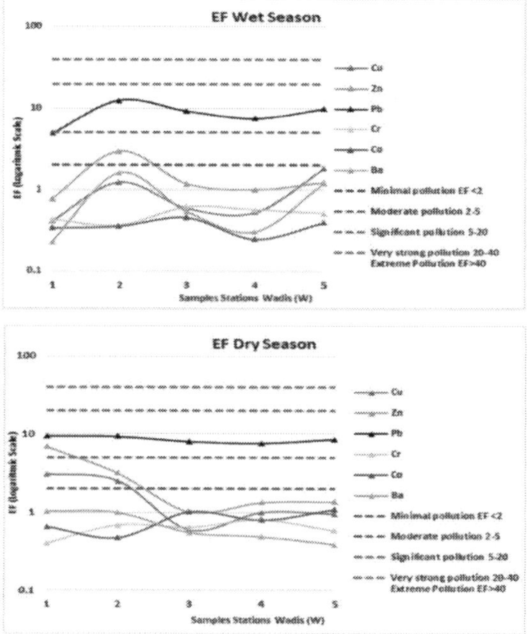

Fig. 2. Wet and dry enrichment factor.

In fact, the Marchica watershed drains rainwater from its four morphological structures by conveying wastewater from various industrial and mining activities through the main wadis that reach the Marchica lagoon, as reported by ABHM (2009) [7]. Some industries with main activities related to the steel industry were installed before implementing the national law on environmental impact studies and the discharge of their industrial waters without prior treatment, as reported by LNESP /BBN (2016) [14]. The irrigated agriculture of the Bouarg plain often uses excessive quantities of fertilizers, as reported by Re et al. (2014) [15] and Snoussi (2004) [16]. The iron ore deposit, part of which has been intensively exploited and

https://doi.org/10.1051/e3sconf/202129803001

abandoned without rehabilitation, can generate the production of acid mine drainage and the release of trace elements according to Bussière et al. (2010) [17] and impacts the Marchica lagoon through the Cabaillo wadi as reported by Khafouri and Talbi (2020) [18] and Lakrim et al. (2016) [19].

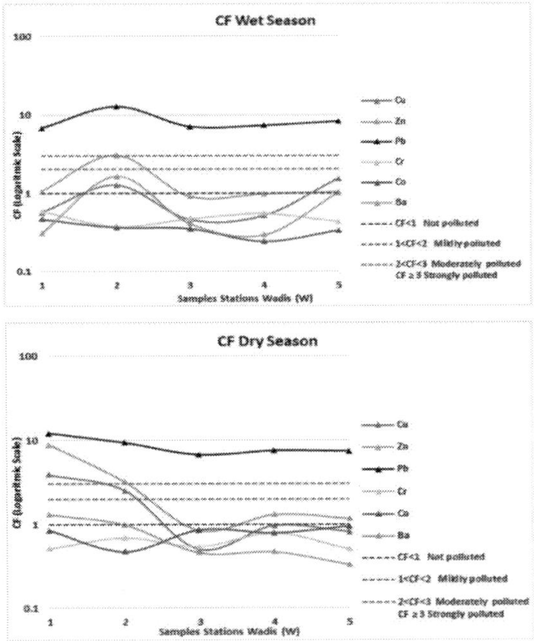

Fig. 3. Wet and dry contamination factor.

Similar pollution was found in the sediments Marchica lagoon during the same period of this study, as reported by Oujidi et al. (2021) [8]. This can confirm that the watershed has negatively impacted the Marchica lagoon despite the restoration actions around this Ramsar site during the last decade.

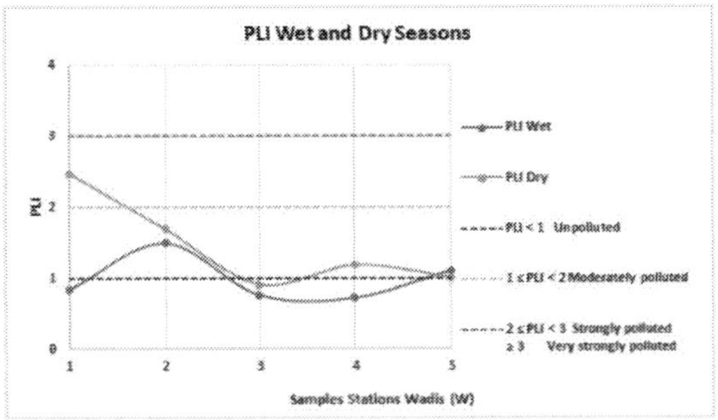

Fig. 4. Wet and dry pollution-load index.

6 Conclusion

This study confirms through the ecotoxicological risk assessment of the trace elements using pollution indices and SQGs, contamination of Marchica watershed by Pb, Zn, and Cu. Therefore, an environmental management strategy is necessary, taking into account the rehabilitation of the abandoned mining sites, the rational use of pesticides and fertilizers in agriculture, and the environmental responsibility of industrial companies, especially those installed before the implementation of the national law on environmental impact studies.

Acknowledgments

The author thanks all partners of the Lag-Nad 2016 project who supported this work

References

1. M. Kennish and H. Paerl, "Coastal Lagoons," Marine Science, pp. 1–15, Jun. (2010).
2. M. J. Kennish, Ed., "Encyclopedia of Estuaries." Encyclopedia of Earth Sciences Series (2016).
3. M. Snoussi et al, "Effects of the construction of dams on the water and sediment fluxes of the Moulouya and the Sebou Rivers, Morocco," Regional Environmental Change, vol. **3**, no. 1–3, pp. 5–12, Dec. (2002).
4. L. B. Korman et al, "Spatially distributed simulations of dry and wet season sediment yields: A case study in the lower Rio Loco watershed, Puerto Rico," Journal of South American Earth Sciences, vol. **103**, p. 102717, Nov (2020).
5. O. Raji et al, Von Grafenstein, and A. Poujol, "Climate and tectonic-driven sedimentary infill of a lagoon as revealed by high resolution seismic and core data (the Nador lagoon, NE Morocco)," Marine Geology, vol. **398**, pp. 99–111, Apr. (2018).
6. M. K. Bloundi et al, "Heavy metal contamination of coastal lagoon sediments by anthropogenic activities: the case of Nador (East Morocco)," Environmental Geology, vol. **56**, no. 5, pp. 833–843, Jan. (2009).
7. ABHM (Agence du basin hydraulique de la Moulouya). Étude de protection du Grand Nador contre l'inondation. Rapport, 150 p. (2009).
8. B. Oujidi et al, "Seasonal and Spatial Patterns of Ecotoxicological Indices of Trace Elements in Superficial Sediments of the Marchica Lagoon Following Restoration Actions during the Last Decade," Diversity, vol. **13**, no. 2, p. 51, Jan. (2021).
9. R. A. Sutherland, "Bed sediment-associated trace metals in an urban stream, Oahu, Hawaii," Environmental Geology, vol. **39**, no. 6, pp. 611–627, Apr. (2000).
10. L. Hakanson, "An ecological risk index for aquatic pollution control.a sedimentological approach," Water Research, vol. **14**, no. 8, pp. 975–1001, Jan. (1980).
11. D. L. Tomlinson et al, "Problems in the assessment of heavy-metal levels in estuaries and the formation of a pollution index," Helgoländer Meeresuntersuchungen, vol. **33**, no. 1–4, pp. 566–575, Mar. (1980).

12. R. L. Rudnick and S. Gao, "Composition of the Continental Crust," Treatise on Geochemistry, pp. 1–64, (2003).
13. D. D. MacDonald et al, "Development and Evaluation of Consensus-Based Sediment Quality Guidelines for Freshwater Ecosystems," Archives of Environmental Contamination and Toxicology, vol. **39**, no. 1, pp. 20–31, Jun. (2000).
14. LNESP (Laboratoire National des Etudes et de Surveillance de la Pollution), Département de l'Environnement. Rapport Mise à jour du Bilan de Base National d'émissions/Rejets de Polluants (BBN). In MEDPOL program for the assessment and control of marine pollution in the Mediterranean, Report 88 pp. (2016).
15. V. Re et al, "Identifying the effects of human pressure on groundwater quality to support water management strategies in coastal regions: A multi-tracer and statistical approach (Bou-Areg region, Morocco)," Science of The Total Environment, vol. **500–501**, pp. 211–223, Dec. (2014).
16. M. Snoussi, "Review of certain basic elements for the assessment of environmental flows in the Lower Moulouya." *IUCN International Union for Conservation of Nature, Gland* (2004).
17. B. BussiÈre, "Acid Mine Drainage From Abandoned Mine Sites: Problematic and Reclamation Approaches," Advances in Environmental Geotechnics, pp. 111–125, (2010).
18. K. Azzeddine and T. El Hassan, "Assessment of Metallic Contamination of Water Resources in the Area Around the Abandoned Mining Site of Ouixane (North East Morocco)," Proceedings of the 4th Edition of International Conference on Geo-IT and Water Resources 2020, Geo-IT and Water Resources 2020, Mar. (2020).
19. M. Lakrim et al, "Application Géomatique Pour La Cartographie De La Vulnérabilité Environnementale Engendrée Par Les Déchets Miniers De La Mine Ferrifere De Nador (Nord-est du Maroc)," European Scientific Journal, ESJ, vol. **12**, no. 15, p. 287, May (2016).

Sheep meat Beni-Guil breed: A local meat labeled PGI to promote quality and preserve biodiversity in Morocco

Kamal Belhaj[1,*], *Farid* Mansouri[1], *Melhaoui* Reda[1], *Yassine* Taaifi[1], *Youssef* Rbah[1], *Marianne* Sindic[2], *Mohamed* Boukharta[3], *Hana* Serghin-Caid[1], and *Ahmed* Elamrani[1]

[1]Laboratory of Agricultural Production Improvement, Biotechnology and Environment, Faculty of Sciences, University Mohammed First, 717, 60000 Oujda, Morocco.
[2]Analysis Quality and Risk Unit, Laboratory of Food Quality and Safety (QSPA), Gembloux Agro-Bio Tech, University of Liège, 5030 Gembloux, Belgium
[3]Institute of Agricultural Industries, High school of Charlemagne, 4500 Huy, Belgium.

Abstract. Currently, meat consumers are more conscious about the authentic meat products linked to their origin, which generates a growing demand for labeled sheep meat. In eastern Morocco, the Beni-Guil (BG) sheep meat is a Protected Geographical Indication (PGI) product. This study aimed to assess the variation of intramuscular fat quality of BG sheep in its cradles (Cr). The chemical composition of the longissimus lumborum muscle was evaluated from 20 female lambs obtained in the two main cradles of the BG breed (Ain Beni-Mathar (BG-Cr1) and Tendrara (BG-Cr2). The results show that the FA profile was affected by Cr ($p<0.05$). The oleic acid, palmitic acid, stearic acid, and linoleic acid are the dominants FA of analyzed meats. The polyunsaturated fatty acids (PUFA)/saturated fatty acids (SFA) and PUFA n-6/PUFAn-3 ratios respectively range from 1.15 and 5.02 for BG-sheep meats from Cr1 to 1.36 and 10.73 for those from Cr2. Overall, the results indicate that the analyzed meats have an interesting FAs profile with favorable ratios and indices. Thus, this study provides new information on the lipid quality of BG-IGP sheep meat produced extensively in two main cradles of this typical Moroccan sheep breed.

1 Introduction

The global demand for safe and high-quality meat has increased significantly in recent years. Consumers attach great importance to the extrinsic meat quality in response to raising safety and health concerns, and consequently, are more conscious about the authentic meat products linked to their origin. This circumstance generates a growing demand for labeled sheep meat. The provenance of meat is essential information for breeders, retailers, and consumers. For the latter, red meat may have a particular value associated with the geographical origin or production system from which they derive, e.g., "Protected Designation of Origin (PDO)", "Protected Geographical Indication (PGI)", and "organic", etc.

The use of local sheep breeds, reared under traditional-local systems and commercialized under the labels of designation of origin, has become very important in recent decades. In Morocco, sheep production is based mainly on pastoral systems with pure local breeds, i.e., Beni-Guil, Timahdite, Sardi, D'man, and Boujaad. The Beni-Guil sheep breed is considered the typical sheep of eastern Morocco and is recognized as the first red meat to be regarded as a PGI in Morocco. This sheep breed is born, reared, and slaughtered in the eastern Moroccan region, fed a particular diet, and reared under a specific pastoral-transhumant system. Consequently, it has a unique sensory and nutritional quality meat linked to this distinct environment and natural grazing system. Beni-Guil sheep meat is one of the most consumed meat due to its sensory, organoleptic, and attractive nutritional value [1].

These characteristics imply its potential use in a sustainable and/or biological economic system based on the traditional transhumant grazing system. The meat quality is affected by intrinsic and extrinsic factors, which are generally related to a geographical area. Previous studies have reported geographical area (dietary differences between regions) on fatty acids profile and sensory quality [2]. Notably, the flavor is associated with the types of food, where lambs raised under the pasture-based system produce meat with pastoral flavor.

This paper aims to study the variation of intramuscular fat quality of PGI-Beni-Guil local sheep breeds according to in its main cradles. To date, no available studies have looked at the effects of the cradle on the meat quality characteristics of the Beni-Guil sheep breed. This work is a continuation of the one started in 2016-2017 to confirm the effect of the cradle on the Beni-Guil meat quality [3].

* Corresponding author: belhaj.kamal90@gmail.com

2 Material and methods

2.1 Animals

Beni-Guil sheep are distributed over the eastern region of Morocco. It is well suited to the harsh climate and poor pasture conditions and is classified as a meat production breed. It is a breed characterized by fire brown head, ears, prothorax, and legs without white spots; white fleece; absence of horns in the female (Fig. 1). The rams usually have coarse and longhorns that are pale and spiral. Animals were reared under transhumant-local production systems in the pasture of eastern Morocco highlands. This study was done on twenty-four 7-month-old of Beni-Guil female lambs weaned at the age of 3 months. The samples were taken from the two main cradles of the Beni-Guil breed: Female lambs of the Beni-Guil breed reared and slaughtered in Ain Beni-Mathar (Cr1; n=12) and Tendrara (Cr2; n=12). The pre-determined live weight of studied animals was 35.67 kg for Cr1 and 34.48 kg for Cr2. Female lambs were slaughtered in their areas of origin in the slaughterhouse.

Further details about rearing, diet composition, and slaughter conditions are described by [4]. The sampling was carried out in spring 2018. The *longissimus lumborum* samples were taken 24 hours *post-mortem* with the assistance of a slaughterhouse veterinarian.

Fig 1. Beni-Guil female lambs (18 months old) in Ain Beni-Mathar region (A) and Tendrara region (B)

2.2 Fatty acids composition

The fat was extracted according to the Bligh Dyer method [5]. The fatty acids were methylated, and the samples were analyzed by Gas Chromatography equipped with a flame-ionization detection and capillary column. The nonadecanoic acid 1 mg/ml was used as an internal standard to calculate the fatty acid concentration.

2.3 Statistical Analyses

All the data were presented as means±standard deviation of triplicate determinations. The statistical analyses were achieved using the software SPSS version 20 and considering the cradle (Ain Beni-Mathar cradle vs. Tendrera cradle) as the source of variation. The means were compared using the *t-test,* and the significance was declared at $p<0.05$.

3 Results and discussion

The prevalent fatty acid content, ratios, and lipid health indices of the Beni-Guil sheep meat are shown in Table 1. The results show that the cradle has a significant effect on the fat quality of BG-PGI sheep meat ($p<0.05$). The results show no significant differences between the studied groups for the dominant saturated fatty acids (Palmitic: 2507,03 vs. 2652,14 and stearic acid: 1381,26 vs. 1426,97 mg/kg of muscle, respectively for Cr1 and Cr2) and Monounsaturated fatty acid (oleic acid: 3735,01 vs. 3707,43 mg/kg of muscle).

However, female lambs from Cr2 showed significantly higher value for polyunsaturated fatty acids than Cr1 ($p\leq0.001$). On the other hand, no significant difference between Cr1 vs. Cr2 was found for the SFA and MUFA sums. Concerning lipid health ratios and indices, the meat from Cr2 represents the better ratio (PUFA/SFA: 0.93 vs. 0.83) and indices (Thrombogenic: 0.93 vs. 0.83 and Atherogenic index: 0.82 vs. 0.69). In contrast, the meat of Cr1 has a better PUFA-n-6/PUFAn-3 ratio (5.02 vs. 10.73). In the present study, oleic (C18:1), palmitic (C16:0), stearic (C18:0), and linoleic acid (C18:2n-6) represent more than 80% of the total fatty acid concentration of the *longissimus lumborum* of Beni-Guil sheep meat.

The dominant FA in the obtained profile was C18:1 in the meat of two studied groups, followed by C16:0 and C18:0 (Table 1). These results align with those reported by several authors for sheepmeat [4, 6]. To fight against certain diseases linked to the current lifestyle, the ratio of PUFA/SFA recommended should be greater than 0.4 and that of n-6/n-3 between 1 and 4. These two ratios are used as a common index to assess lipid nutritional quality. The omega-6/omega-3 ratio plays an essential role, in particular in the prevention of cardiovascular diseases [7]. The results of this study show that the cradle affects these ratios significantly. The muscle of BG-Cr2 contains more PUFA resulting in the higher PUFA/SFA ratio (1.15 vs. 1.36), while the BG-Cr1 one shows a lower n-6/n-3 (5.02 vs. 10.73). The individual effect of fatty acid on cholesterol metabolism was evaluated by calculating the Hypocholesterolemic (h)/Hypercholesteremic (H) ratio. The results demonstrate that the meat of female lambs reared in Cr2 has the better h/H ratio (1.72 vs 2).

This finding could be explained by the breeding system and the pasture availability and diversity between studied cradles. In the Cr2, the breeding system is more extensive than Cr1, where the *Artemisia herba-alba* and *Stipa tenacissima* formations are in an advanced degradation state compared to Cr2 [8]. Consequently, the farmers of Cr1 adopt a semi-extensive system using barley and Lucerne hay as a food supplementation to satisfy the different needs of their herds. The wormwood and alfa species are richer in bioactive substances such as antimicrobials and antioxidants, which lead to the decrease in ruminal bio-hydrogenation and PUFA degradation [9]. As a result, the production of meat rich in PUFA is the case of meat from Cr2. Concerning the PUFA-6/PUFA-3 ratio, the recorded results are higher than the recommended value. The incorporation of alfalfa could explain the lower value recorded in meat produced in Cr1 compared to Cr2 in the food ratio rich in n-3 fatty acids.

Table 1. Fatty acids content of Beni-Guil sheep meat in Ain Beni-Mathar and Tendrara region

	Cradle		
Prevalent FAs (mgkg^{-1} of muscle), **sums, ratios, and indices**	**BG-Cr1**	**BG-Cr2**	**P-value**
Oleic acid	3735,01±660,40	3707,43±517,26	0.580
Palmitic acid	2507,03±436,01	2652,14±201,68	0.903
Stearic acid	1381,26±308,77	1426,97±151,74	0.289
linoleic acid	737,62±230,91	1235,60±191,82	<0.001
Arachidonic acid	208,36±98,53	697,24±280,83	<0.001
Linolenic acid	90,15±24,10	116,26±47,99	0.1
Docosahexaenoic Acid	83,24±17,73	76,21±16,58	0.278
SFA	4658,38±732,67	4978,27±338,75	0.167
MUFA	4064,64±704,35	4192,55±413,29	0.572
PUFA	1286,84±394,74	2552,02±520,80	<0.001
UFA	5351,49±822,38	6744,57±613,95	<0.001
PUFA/SFA	1,15±0,078	1,36±0,08	<0.001
ω-6/ω-3	5.02±1.74^{a}	10.73±4.38^{b}	0.001
TI	0.93±0.11^{c}	0.83±0.04^{b}	0.002
AI	0.82±0.15^{c}	0.69±0.01^{cb}	0.002
h/H	1.72±0.26^{a}	2±0.06^{b}	<0.001

FAs: Fatty Acids; **SFA**: Saturated Fatty Acids; **UFA**: unsaturated Fatty Acids; **PUFA**: polyunsaturated Fatty Acids; **TI** = Thrombogenic Index: [C14 :0 + C16 :0] / [(0.5*MUFA) + (0.5*∑n−6) +(3*∑n−3) + (n−3 / n−6)]. **AI** = Atherogenic Index : [(4*C14 :0) + C16 :0] / [(PUFA) + (MUFA)], without the inclusion of 18:0 which is considered to be neutral on serum cholesterol; Hypocholesterolemic **(h)**/Hypercholesteremic **(H)** = (18 :1n9c + 18 :2ω6 + 20 :4ω6 + 18 :3n3 + 20 :5n3 + 22 :6n3) /(14 :0 + 16 :0).

Conclusion

The present research allowed us to study the variation of sheep meat quality from the "Beni-Guil-PGI breed" in its main cradles. The results show that the lipid nutritional value of this labeled breed is not homogeneous. To increase consumers' loyalty, the PGI label should consider the cradle effect on meat quality in their regulatory directive to provide a standard product from one purchase to another. Further investigations are needed to study the cradle effects on meat traits, evaluating, in particular, the sensorial quality to identify the olfactory biomarkers for each cradle.

References

1. K. Belhaj, F. Mansouri, M. Sindic, M.-L. Fauconnier, M. Boukharta, H. Serghini Caid, and A. Elamrani, The Scientific World Journal, **2021**, (2021)
2. S.W. Erasmus, M. Muller, and L.C. Hoffman, Journal of the Science of Food and Agriculture, **97**, 1979-1996 (2017)
3. K. Belhaj, F. Mansouri, A. Benmoumen, M. Sindic, M.-L. Fauconnier, M. Boukharta, C.H. Serghini, and A. Elamrani, Archives Animal Breeding, **63**, 471-482 (2020)
4. K. Belhaj, F. Mansouri, M. Sindic, M.-L. Fauconnier, M. Boukharta, H. Serghini Caid, and A. Elamrani, Journal of Food Quality, **2021**, (2021)
5. E.G. Bligh and W.J. Dyer, Canadian journal of biochemistry and physiology, **37**, 911-917 (1959)
6. M. Blasco, M.M. Campo, J. Balado, and C. Sañudo, Journal of the Science of Food and Agriculture, **99**, 3335-3342 (2019)
7. NL. Howes, AEDA Bekhit, D.J. Burritt, and A.W. Campbell, Comprehensive Reviews in Food Science and Food Safety, **14**, 22-36 (2015)
8. A. Bechchari, A. El Aich, H. Mahyou, B. Baghdad, and M. Bendaou, Journal of Materials and Environmental Science, **5**, 2572-2583 (2014)
9. V. Vasta, D. Aouadi, D.M. Brogna, M. Scerra, G. Luciano, A. Priolo, and H.B. Salem, Meat science, **95**, 235-241 (2013)

Physico-chemical characterization of water and metallic contamination of Mytilus galloprovincialis in the coastal zone of Saidia

Khaoula Kasmi[1], *Kamal* Belhaj[1], and *Abdelhafid* Chafi[1]

[1]Laboratory of Agricultural Production Improvement, Biotechnology and Environment, Faculty of Sciences, University Mohammed First, 717, 60000 Oujda, Morocco.

Abstract. Water is a vital natural resource for life. The metal pollution of aquatic ecosystems is increasing due to the effects of urbanization and industrialization. As a result, heavy metals contamination has become a serious threat to public health because of their toxicity, long persistence, bioaccumulation, and biomagnification in the food chain. This investigation aims to evaluate seawater's Physico-chemical quality and assess the contamination risks of Mytilus galloprovincialis mussels by Cadmium, Zinc, Iron, and Copper. The study was carried out in three stations of Saidia coast during July 2019. The seawater was analyzed for its Physico-chemical parameters, and ICP-AES determined the heavy metals content in mussels. The results show that the tested water has a pH value ranging between 7.48-7.62 with nitrate and nitrite content range between 0-4.78 mg/L, 0.007-0.181 mg/L, respectively. The recorded water conductivity and oxydability values ranged from 22 to 55 mS/cm and from 5.67 to 10.47 mg/L. The mean concentration found in the mussels were in decreasing order, Iron (13.5 mg/kg), Zinc (2.3 mg/kg), Copper (0.65 mg/kg) and Cadmium (0.02 mg/kg), respectively. The obtained results indicate that all values are below the threshold recommended by Food and agriculture organization and world health organization.

1 Introduction

Marine pollution is one of today's most current concerns. The metal pollution of aquatic ecosystems is increasing due to urbanization and industrialization (anthropogenic activities). The sources of pollution in the marine environment include the paint on ships, diverse industrial processes, and agricultural chemical products. Land undergoes high development pressure; consequently, a large contaminant load enters the water system and causes water quality degradation. As a result, and over time, the contamination by heavy metals occurs, it is recalled that these latter are hazardous contaminants and consist a real threat to public health because of their toxicity, long persistence, bioaccumulation, and biomagnification in the food chain and the various segments of the environment such as water sediments and biota. Fish is a highly appreciated food and is consumed all over the world for its nutritional value. This species is known for its nutritional and therapeutic benefits. In addition to its essential high-protein quality, fish typically have rich contents of oligo-element, vitamins, and long-chain fatty acids n-3; Globally, fish accounts for about 17 percent of animal protein intake and almost 6% of proteins consumed by human beings (Jose et al., 2007).

However, fish could become a veritable threat to human health because of the heavy metals' accumulation via marine water. The fish in the aquatic system has several assimilation methods, including organ bioaccumulation (liver, heart, and kidney, etc.), ion exchange, and tissue absorption. This investigation aimed to evaluate seawater's Physico-chemical quality and assess the contamination risks of *Mytilus galloprovincialis* mussels by Cadmium, Zinc, Iron, and Copper.

2 Material and methods

The study was carried out in the summer season. Water samples were taken from three stations (S): **S1**: Saidia coast at the Algerian-Moroccan border near Oued kiss; **S2**: Saidia coastal zone; **S3**: near the mouth of the Moulouya. On the other hand, the *Mytilus galloprovincialis* mussel samples were only taken from S1 since they are absent in the two others (absence of rocks) (Fig. 1).

The Physico-chemical analyses of water were carried out at the Oriental Center for Water Sciences and Technologies (COSTE). The analyzed parameters were the pH, oxidability, electrical conductivity, nitrites, and nitrates concentrations. The samples of *Mytilus gallonprovincialis* were cleaned by removing algae and debris using distilled water and were shelled for subsequent analyses. One gram of dried mussel was used for the elements metals analysis. The samples' mineralization was carried out by adding concentrated nitric acid (1 ml/100 mg of dried matter).

* Corresponding author: kasmikhaoula77@gmail.com

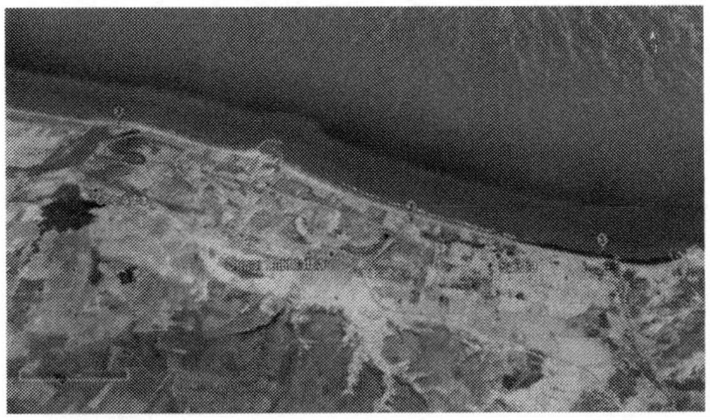

Fig 1. The sampling stations area

3 Results and discussion

The results of the Physico-chemical analysis of studied seawater are presented in table 1. The pH values of the seawater measured in different stations 1, 2, and 3 are between 7.62 and 7.48. These values are lower than the pH values of Mediterranean seawater (8,074 ± 0,034) [1]. This result could be explained by the geochemistry of the area while suspecting the existence of agro-industrialist effluents on both sides of the Moulouya and the Kiss wadi. In addition to the tributaries discharged into the Moulouya carrying a significant amount of organic material, which, while degrading, generates substances in its acidogenic phase that causes the pH decrease of the seawater. Similar results were obtained by Hassoun [1]. Karol, Bernd [2] reported that organic waste causes a decrease in the pH of seawater by the formation of bases via the release of protons.

Table 1. Physicochemical analysis of Mediterranean seawater from the three studied stations

Parameters	Station 1	Station 2	Station 3
pH	7.62±0.03	7.47±0.01	7.48±0.15
Oxydability (mg O_2L^{-1})	5.62±1.01^a	9.3±2.13^b	10.47±2.74^c
Conductivity (mS/cm)	55±10.13^a	55±8.9^a	22±3.6^b
Nitrate (mg/L)	ND	0.87^b	4.78^c
Nitrite (mg/L)	0.007±0.00^a	0.013±0.00^b	0.181±0.00^b

Values with different letters are significantly (p<0.05) different within rows; **ND**: not detected

Table 2. Trace element content in Mytilus gallonprovincialis mussels from Station 1

Trace element (mg/kg)	Mean±SD	Standard (mg/kg)
Cadmium	0.02±**0.00**	1
Copper	0.65±**0.00**	50-150
Zinc	2.3±**0.17**	200-500
Iron	13.5±**1.43**	100-150

3.1 The nitrite and nitrate content:

The values found for nitrates range between 0 mg/L and 4.78 mg/L, respectively, for S1 and S3. As for the nitrites' values range between 0.007 mg/L and 0.18 mg/L, respectively, for S1 and S3 (Table 1). The nitrite values recorded are low due to its oxidation to nitrate via the phytoplankton or by photo-reduction.

The increase in nitrate concentration in the third station located at the mouth of the Moulouya could be due to the contributions received by Oued Moulouya, which is experiencing severe degradation, especially in agricultural activity in the region. In addition, the intensive use of nitrogenous fertilizers and irrigation techniques pollute the waters of Moulouya. This water's flow loaded with nitrate towards the sea could cause this increase in nitrate at the third station.

3.2 Variation in Oxidability

The obtained oxidability values range between 5,62mg of O2/L and 10,47 mg of O2/L, respectively, for S1 and S3 (Table 1). The stations' oxidability values show that the further we go towards Moulouya, the more the value of the oxidability increases; That is to say, the load of oxidizable matter increases in the middle. The oxidability recorded at the second station is very close to the third; this is explained by the short distance between the two stations.

Therefore, we could assume that the origin of the oxidizable matter present in the zone is the contributions coming from Moulouya. This river knows a substantial degradation due to agricultural activities on the level of the Triffa plain, as well as domestic and industrial plants in the towns of Missour, Outat El-Haj, Tindit, Guercif, and Taourirt, which liquid discharges directly without prior treatment, which influences the entire watershed of the Moulouya [3].

3.3 The variation of electrical conductivity

The recorded values for EC range between 55 mS/cm for both S1 and S2 and 22 mS/cm for S3 (Table 1). From the results, we notice that the conductivity of the two stations S1 and S2 is in the norm, which is 56 mS/cm. The increase in the conductivity of a solution is due to the intense concentration of salts such as chlorides [1]. On the other hand, the value recorded in S3 is lower than the norm (22 mS/cm). This decrease would be due to the dilution of the resort's waters by the fresh waters of Moulouya. Large inflows of fresh water, resulting in a decrease in the mean salinity of the water [4]. Agricultural activity in the Triffa plain would probably have contributed to the decrease in the conductivity of the water from the third station.

The influence of agricultural activities on many Mediterranean wetlands has often disrupted their functioning. Tamisier and Grillas [4] have reported that many temporary systems have been converted to permanent systems, and salinity has decreased due to agricultural production flows, which often dominate diets of salinity.

3.4 Heavy metals

The results of the heavy metals concentrations (Cd, Cu, Zn, and Fe) in the Mussels collected from the Saidia coastal zone are given in Table 2. The mean concentration in the mussel is the following sequence Fe > Zn > Cu > Cd during summer. The copper content found in mussels is around 0.65 mg/kg. This value is very far from the limits set by the FAO (50 to 150 mg/kg); Likewise, if we compare it with the limit value determined by Nauen [5], which is 30 µg/g. However, our result is in agreement with those reported by Schintu, Durante [6] which (5-13 mg/kg) and remain lower than those reported in the literature for other species of mollusks bivalves: 38mg/kg for *Chamelea.gallina* [7] and 175mg/kg for the species *Donax.trunculus*. The Cd content in mussels is in the order of 0.02 mg/kg dry matter. This registered value is lower than the European Community permissible level of Cd for human consumption (1 mg/kg wet weight). In addition, our result is below the limit fixed by the Economic Commission for Europe of the order of 1 mg /kg. Zinc appears to have a protective effect against the toxicities of cadmium and lead. However, in excess quantities, essential elements can also be toxic and cause serious health threats to human. The recorded value is of 2.3 mg/kg and is considerably lower than the limits set by FAO (200 to 500 mg/kg). Concerning the iron content in the sample of mussels taken is of the order of 13.5 mg/kg. The Fe content is lower than those found in the literature for the same species of the order of 2462 mg/kg [8] and other species of bivalves such as *Scrabicularia.plana* (1657-3-4452mg/kg) [9], *Cardium.edule* (950-1026 mg/kg) [10] and *Mytilus edulis* (1380 mg/kg) [11].

4 Conclusion

Physico-chemical characterization of water and metallic contamination of *Mytilus galloprovincialis* in the coastal zone of Saidia has been studied. The assessment of the seawater physicochemical parameters at the level of the Saidia coastal zone allowed us to deduce the negative effect of agricultural intensification in the Triffa plain on seawater quality. However, the result of heavy metal content shows that all the recorded values are acceptable. Further investigations are needed to study the Eco-toxicological risks of Saidia coastal to determine the different pollutants and their origins to establish corrective measures for sustainable development.

References

1. A.E.R. Hassoun, *Analyse et Modélisation de l'Acidification en Mer Méditerranée*, in *Énergie et Environnement "Institut de Modélisation et d'Analyses en GéoEnvironnement et Santé"*. 2014, Université de perpignan via domitia: Perpignan. p. 216.
2. K. Karol, S. Bernd, H. Karoline, U. Machulik, and D. Schulz-Bull, Journal of Marine Systems, **132** 106–115 (2014)
3. M. Makhoukh, M. Sbaa, A. Berrahou, and M. Van Clooster, LARHYSS Journal P-ISSN 1112-3680/E-ISSN 2521-9782, (2011)
4. A. Tamisier and P. Grillas, Biological Conservation, **70**, 39-47 (1994)
5. C.E. Nauen, FAO Fisheries Circular (FAO). no. 764., (1983)
6. M. Schintu, L. Durante, A. Maccioni, P. Meloni, S. Degetto, and A. Contu, Marine Pollution Bulletin, **57**, 832-837 (2008)
7. J. Usero, J. Morillo, and I. Gracia, Chemosphere, **59**, 1175-1181 (2005)
8. F. Çevik, M.Z.L. Göksu, O.B. Derici, and Ö. Fındık, Environmental monitoring and assessment, **152**, 309-317 (2009)
9. Cheggour and M., *Contribution à l'étude d'un milieu paralique : l'estuaire de Bou regreg (Côte atlantique marocaine) : conditions écologiques globales ; étude de la contamination métallique*, in *Biologie*. 1998, Université Mohamed 5: Rabat. p. 337.
10. A. Kaimoussi, A. Mouzdahir, S. Bakkas, And A. Chafik, Bulletin de l'Institut Scientifique, Rabat, **1997**, 95-102 (1998)
11. J. Pempkowiak, A. Sikora, and E. Biernacka, Chemosphere, **39**, 313-321 (1999)

E3S Web of Conferences **298**, 03004 (2021)
ICCR 2020
https://doi.org/10.1051/e3sconf/202129803004

Carob tree micropropagation essays (Ceratonia siliqua L.)

Sara Nia[1,*], *Malika* Abid[1], and *Ilham* Belkoura[2]

[1]Department of biology. FS Oujda, Morocco
[2]Department of basic sciences. National School of Agriculture, Meknes, Morocco

Abstract. Carob tree (*Ceratonia siliqua L.)* is an agro-sylvo-pastoral species that has been classified among the most efficient trees. The first publications on the "*in-vitro*" multiplication of Carob are very recent and show conflicting results. Mastery of this technique seems necessary to provide the increased demand of the international market for the gum extracted from the endosperm of the seed, a polyoside called galactomannan. This work aimed to study micropropagation from adult tree micro cuttings. Our results showed that the most favorable disinfectant is $HgCl_2$ with 93% surface sterile explants. Comparison of the effect of various hormones with different concentrations (BAP, Zeatin, GA3, TDZ, and ANA) has shown that the addition of BAP on MS medium gave a good response of axillary bud development (58,3%±6,42) while BAP supplemented with activated charcoal gave the best results (87,3%±4,23). As for rooting, the different alternatives of hormones types and concentrations must be later considered to initiate rooting of the buds.

1 Introduction

Mediterranean ecosystems suffer from scarce or irregular rainfall and long dry summer periods combined with high anthropogenic pressure, leading to irreversible degradation of the vegetation cover and rapid soil erosion. Faced with this critical situation, reforestation becomes an absolute necessity to conserve soil fertility and improve the standard of living of the rural population by using pioneer multipurpose tree species adapted to climatic hazards and marginal and eroded land the carob tree [1,2]. In addition to its rusticity, its indifference towards the nature of the soil, its quality wood, its ornamental and landscape value, this species acquires an interest increasingly growing at the level of Morocco, because of its seeds whose industrialization knows a remarkable development and of the commercial transactions with a value which exceeds by far that of the ligneous production. Thus, the pulp, seeds, and gum are subject to an essential trade towards Europe. Moreover, previous studies showed that the Moroccan cultivars are characterized by a high yield of seeds that provide a high yield of the endosperm.

As a result, the kingdom is constantly improving its products to ensure a good supply to the local industry to extract gum [1,3]. Given the importance given to the carob tree in recent years, traditional propagation techniques could not satisfy the increased need for this plant. Therefore, the use of in vitro culture represents a significant challenge and a promising solution [4]. Therefore, the purpose of this study was to shine new light on these debates and to optimize reliable methods for rapid propagation of this species with a focus on micropropagation by micro-cutting

2 Material and Methods

2.1 Sterilization and establishment of plant material

Shoots of the current year or young offshoots were collected during different periods: September, February, and May, from the same adult female tree (under rain-fed conditions in Meknes), to surmount the genotypic effect.

Explants were defoliated just after collection and surface disinfected with running water and Tween 80 to remove surface impurities, shoots were placed for 2min in ethanol 70%, and later two disinfectants were tried (inside the laminar airflow cabinet) in order to control contaminations: HgCl2 (0,5g/l, 1g/l, 2g/l for 3 or 5min) and commercial bleach solution –NaClO (12°)-for different times (10, 15, 20 and 30min), and were washed with SDW 3 times.

2.2 Axillary bud induction

Only one primary nutrient medium was used for disinfection tests and consisted of MS medium supplemented with sucrose (20 g/l) as a carbon source. Then, the medium was solidified with agar (8g/l).

The compilation of data was done after four weeks, during which time the percentage of sterile surface explants is estimated.

*Corresponding author: niasarah041@gmail.com

2.3 Shoot multiplication

In order to stimulate the response of axillary bud development, the treatments consisted of culturing shoot explants for eight weeks on MS medium supplemented with different phytohormones.
- M1: MS +30g/l sucrose +8g/l agar + BA(0,5mg/l)
- M2: MS +30g/l sucrose +8g/l agar + BA(0,5mg/l) +AC (1 mg/l)
- M3: MS +30g/l sucrose +8g/l agar +ANA(0,5mg/l)+ GA3(0,5mg/l)

The compilation of data was done after eight weeks.

2.4 Rooting induction

The rooting process was separated into two treatments: Shoots were cultured on darkness on:
- M1: MS/2+ sucrose (20g/l) + IBA (0,5mg/l)
- M2: MS/2+ sucrose (20g/l) + IBA (0,5mg/l) +agar (7g/l)
- M3: MS/2 sucrose (20g/l) + ANA (0,5mg/l)
- M4: MS/2+ sucrose (20g/) l+ ANA (0,5mg/l) + agar (7g/l)

After 4 weeks, Shoots were transferred to the basal medium: MS/2 + sucrose (20g/l) + agar (7g/l) and incubated under the same light and temperature regime as shoot multiplication cultures.

3 Results and discussion

3.1 Establishment of Carob axenic cultures

3.1.1 Effect of the disinfectant

In order to control contaminants, two sterilizing agents were tried at various concentrations and different times. Pearson's chi-squared test indicated a statistically significant difference between the two sterilizing agents. $HgCl_2$ is more effective with 93,3% of sterile surface explants than NaClO with only 50% of disinfection, as shown in Table1. This result follows previous studies that showed that $HgCl_2$ treatment for 5 min is an effective decontamination procedure in tissue culture [5].

Table 1. Effect of the disinfectant agents on contamination rate (Pearson's chi-squared test).

	NaClO	**$HgCl_2$**
Contamination rates	50%±9,28	6,7%±4, 63
p-value (Pearson's chi-squared test)	0,000	

3.1.2 Effect of the season

It has been found that carob regeneration processes in producing in vitro plants are more desirable from explants originated from suckers collected in summer. Indeed, the highest percentage of regenerated explants (90%) and the highest number of explants with free contamination (76,7%) were obtained.

3.1.3 Effect of the explant's position on the branch

Statistical analysis indicates that the position of the explants on the branch has a significant effect on the cultural establishment. Results show that apical explants give a better budburst rate (27, 8%). Moreover, young suckers give the best survival rates and surface-free contamination since, with young suckers, more than 98% of disinfection was achieved.

3.2 Axillary Buds development

3.2.1 Effect of plant growth regulators on shoot development of Ceratonia siliqua

Shoot development has reached during the second week of culture before transfer to hormone-rich media. Statistical analysis has shown a significant effect of the media on shoot development rates. Our results indicate that MS supplemented with BAP furthered the development of several buds in the same node. This may be the result of the expression of existing adventitious buds in the same node. Moreover, stems and leaves growing were improved by adding activated charcoal (87,3%±4,23) as this component counteracts browning and prevents tissue necrosis(Table 2).

Similar results have been reported on the efficacy of BAP alone on shoot development [6-9].

Table 2. Effect of plant growth regulators on budburst development

	MS+BAP	**MS+BAP+AC**	**MS+ANA+GA3**
Budburst development	58,3%±6,42	87,3%±4,23	35,1±6,38
p-value (Pearson's chi-squared test)	0,000		

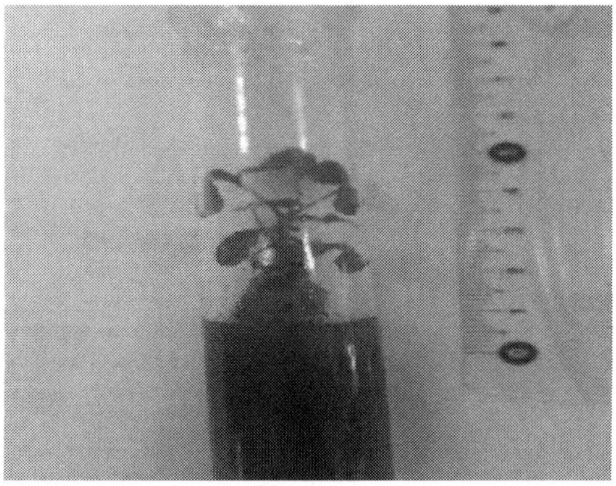

Fig. 1. Shoot-culture at the end of multiplication phase in MS medium

3.2.2 Effect of plant growth regulators on the development of new shoots

Analysis and simulation indicate that the media does not significantly affect the elongation of new shoots. However, the number of leaves per new shoot (Table 3) is high (4,76%±0,29) in the presence of BAP compared to ANA and GA3 (1,33%±0,26).

Table 3. Effect of plant growth regulators on the development of new shoots

	Mean number of new shoots	Length	Mean number of leaves per shoot
ANA+GA3	1,067%±0,06	0,293%± 0,71	1,33%±0,26
BAP	2,233%±0,60	0,793%± 0,24	4,76%±0,29
Significance of one-way ANOVA ($\alpha = 0.05$)	0,061	0,057	0,012

3.3 Rooting

One month after the induction, no rooting was detected. However, the appearance of callus at the base of the explants was considerable, especially in the liquid IBA medium. Although 15 days after transfer to a medium devoid of auxins in the light, we did not note any case of rooting.

The chi-square test demonstrated that the medium has a significant effect on callus production. Moreover, our results indicated that the juvenility of the used shoots has a significant effect on both phenomenons. The absence of rooting in the case of shoots resulting from micro-cutting is undoubtedly due to the age of the shoot, which remains the same age as the initial explant. Therefore, it would be interesting to do a looped micro-cutting first for the intensification of the plant material and then for the rejuvenation of the shoots because the younger the material is, the greater the rooting.

4 Conclusion

This is the first report on in vitro propagation from suckers of the carob tree, which presented a more accessible adaptation to in vitro culture and higher rates of in vitro initialization than explants from shoots of the branch. *In vitro* rooting of Carob shoots are unsuccessful, and different alternative hormones types and concentrations must be later considered to initiate rooting of the buds.

References

1. Ait Chitt M., Belmir H et Lazrak A., 2007. Production de plants sélectionnés et greffés de caroubier. Transfert de technologie en agriculture. Maroc. N° 153, 1-4 pp.
2. Benmahioul B., Kaid-Harche M et Daguin F., 2011. Le caroubier, une espèce Méditerranéenne à usages multiples. In Forêt méditerranéenne, N° 1, 51-58 pp.
3. Battle I. & Tous J., (1997). Carob tree. Ceratonia siliqua L. Promoting the conservation and use of underutilized and neglected crops. 17. Institute of Plant Genetic and Crops Plant Research. Gatersleben/International Plant Resources Institute. Rome. Italy. 92 p.
4. Naoual Gharnit & Abdeslam Ennabili (2015): Categories of Carob Tree (Ceratonia siliqua L.) from Morocco, International Journal of Fruit Science, DOI: 10.1080/15538362.2015.1102674
5. A. Romano, S. Barros, M.A, Martins-Loucao,M.A.(2002). Plant Cell Tiss. Organ Cult. 68, 35–41.(2002).
6. T. Hsina ,N. Elmtili . Cahiers UAE, 7-12. .(2012).
7. R. Saïdi , B. El bouzdoudi , M B.Kbiach, A. Lamarti , A. J. Mater. Environ. Sci, 6(8), 2330-2337. (2015).
8. Shahzad, R. Akhtar , NA. Bukhari . K. Trees:1–15. (2017).
9. El Bouzdoudi , R. Saïdi, Z.N. El Ansari, M.L. El Kbiach, P. Martin, A. Badoc, A.Lamarti. American Journal of Plant Sciences, 8, 2180-2195. (2017).

Studies on riparian communities in eastern Morocco: coastal dayas at the mouth of the Moulouya River

Naoual Zitouni[1,*] , and *Guy* Chavanon[1]

[1]« Laboratoire d'Amélioration de la Production Agricole, Biotechnologie et Environnement », Faculty of Sciences, Mohammed Ist University, Oujda, Morocco

Abstract. This paper aims to study the riparian community of coastal dayas at the mouth of the Moulouya River in the northeast of Morocco. We carried out a monthly sampling of riparian fauna on two dayas (EMDI & EMDII) on the Mediterranean coast. Banks of these dayas are an ecotone where terrestrial, riparian, and aquatic species mix. Riparian species dominate the population in richness and abundance: most species are either sporadic (EMDI) or accidental (EMDII). The dominant species are halophilous. Interesting species have been collected from these dayas: *Gonocephalum yelamosi*, new species for Morocco, *Blaps nitens*, *Dyschirius tensicollis*, and *Dyschirius africanus*, which dayas is a new distribution in the region. Species with the same ecological niches that can compete with each other seem to share the occupation of this environment. The same phenomenon seems to occur for species of the genus *Pogonus* and *Pogonistes*, which have populations that reach their maximum expansion at different periods.

1 Introduction

Riparian communities are made up of species that colonize the wet banks of aquatic environments (ocean, river, torrent, the body of water, dayas, etc.) whatever its size and as long as it is not dried up.

They are colonized by a diversified population, a mixture of terrestrial, riparian, and aquatic origins. They are highly migratory species that can leave the environment as soon as conditions become unfavorable and recolonize others just as quickly. These communities depend on their environment and have rigorous ecological and ethological requirements, making them very originals. They are a diversified mixture of species of terrestrial, riparian, and aquatic origins. The temporary nature and instability of the colonize environments have led them to adapt both ethologically, ecomorphological, and ecophysiologically. This gives them certain originality.

The limited Spatio-temporal position of these habitats allows a comprehensive study of the functioning of their communities. The results thus obtained can easily be extrapolated to more significant situations and applied to more general ecological theories [1].

This particular environment will be the subject of this study. It aims to contribute to the knowledge of riparian environments in general and coastal environments in particular.

2 Materials and methods

2.1 Study area

The two dayas studied are temporary coastal water bodies fed mainly by rainwater (EMDI and EMDII) and groundwater. They dry up in summer (EMDI). They are located near the mouth of the Moulouya (Fig. 1).

Embouchure's Daya I (EMDI) is 200m long and 70m wide and is relatively deep. Its water is salty with a neutral pH, and the substrate is sandy to sandy-silt with a neutral pH. It appears in winter (January) and dries up in summer (July). The riparian vegetation consists mainly of *Salicornia sp*. The banks are narrow and sloping at the beginning and become more comprehensive as it dries out.

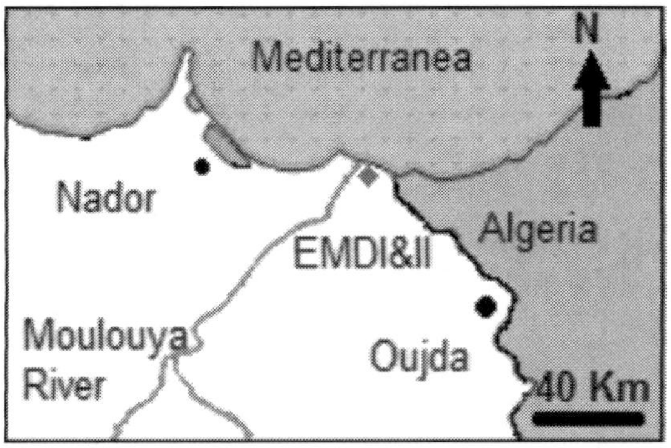

Fig. 1. The geographical location of the study area

* Corresponding author: zitouninaoual@yahoo.fr

https://doi.org/10.1051/e3sconf/202129803005

Embouchure's Daya II (EMDII) is a very shallow daya, which reaches a maximum length of 90m and a width of 35m (January) and dries out in late spring (May). It has salty water with a neutral pH and a silty to a sandy-silty substrate with a neutral pH. Its banks are affluent in organic matter and plant debris.

2.2 Sampling

Surveys were carried out monthly for two consecutive years (two sampling campaigns).

Harvesting was carried out by direct sight hunting during a one-hour active search [2]. All shelters were surveyed, and all fauna found on the banks were collected.

The identifications of Caraboidea were carried out using Antoine's keys [3-7], the other groups were submitted to several specialists.

3 Results and discussion

Thirty-four species were collected on the banks of the coastal dayas of the Embouchure, 27 species on EMDI (Table 1), and 18 species on EMDII (Table 2).

Table 1. Species recorded in EMDI

	Abundance
CARABOIDEA	
Tachys scutellaris dimidiatus	198
Pogonus chalceus	171
Pogonus gilvipes	143
Pogonistes gracilis	68
Syrdenus grayi	49
Emphanes normanus	34
Notaphus ephippium	24
Trichis maculata	12
Cicindela maura	10
Cephalota litorea goudoti	7
Dichirotrichus obsoletus	7
Pogonus luridipennis flavipennis	6
Dyschirus dalila	4
Lophyridia lunulata	3
Dyschirus clypeatus	2
Daptus vittatus	2
Microlestes luctuosus	1
Emphanes lais	1
TENEBRIONIDAE	
Blaps nitens	1
STAPHYLINIDAE	
Bledius bicornis	11
Bledius (Bledius) unicornis	7
Bledius (Euceratobledius) furcatus	4
Bledius (Bledius) angustus	1
Aloconota gregaria	1
DYTISCIDAE	
Agabus didymus	2
Nebrioporus ceresyi	1
HYDROPHILIDAE	
Helochares lividus	4

A new species (*Gonocephalum yelamosi*) for Morocco has been collected from these dayas. In Morocco, they also represent a new station for *Blaps nitens* and a new station, in the region, for *Dyschirius tensicollis* and *Dyschirius africanus*.

Table 2. Species recorded in EMDII

	Abundance
CARABOIDEA	
Dichirotrichus obsoletus	22
Pogonus gilvipes	12
Syrdenus grayi	10
Dyschirus numidicus	10
Tachys scutellaris dimidiatus	9
Lophyridia lunulata	4
Trichis maculata	2
Cicindela maura	2
Dyschirus clypeatus	1
Dyschirus africanus	1
Dyschirus tensicollis	1
STAPHILINIDAE	
Bledius (hesperophillus) debilis	41
Bledius (Euceratobledius) furcatus	2
Platysthetus nitens	1
TENEBRIONIDAE	
Gonocephalum yelamosi	1
Pachychila haroldi	1
Blaps nitens	1
DYTISCIDAE	
Nebrioporus ceresyi	2

The *Caraboidea*, with *Trechidae* and *Scaritidae* families, represent the main component of this type of environment, as in the case of the Spanish salt lagoons [8, 9].

Riparian species dominate the community in terms of both richness and abundance. They have an abundance of 99% (EMDI) to 93% (EMDII) of riparian species with 22 (78%) and 14 (75%) species, respectively. The remainder is distributed between species of terrestrial or aquatic origin.

The majority of species are either sporadic (EMDI, frequency<10%) or accidental (EMDII, 25%>frequency>10%) according to Facet's classification [10]. Very few of them are constant, with a frequency >50%, in the environment (table 3). These are generally halophilic species. On the other hand, diversity (according to the H index of Shannon and Wever [10]) reaches more than 3 bits, and the equitability varies between 67.2% (EMDI) and 80.5% (EMDII).

Table 3. Constant species in EMDI et EMDII

	EMDI	**EMDII**
T. scutellaris dimidiatus	+	.
P. chalceus	+	.
P. gracilis	+	.
P. gilvipes	+	.
S. grayi	+	.
D. obsoletus	.	+

https://doi.org/10.1051/e3sconf/202129803005

Furthermore, the stability coefficient (according to Gautier [11]) is 1.65 and 1.35 in EMDI and EMDII stations, which are relatively stable communities. This indicates that the same species contingent dominates these stations and is found in successive samplings.

Despite their low abundance, tiger beetles appear successively in the banks of both dayas (Fig. 2 and 3).

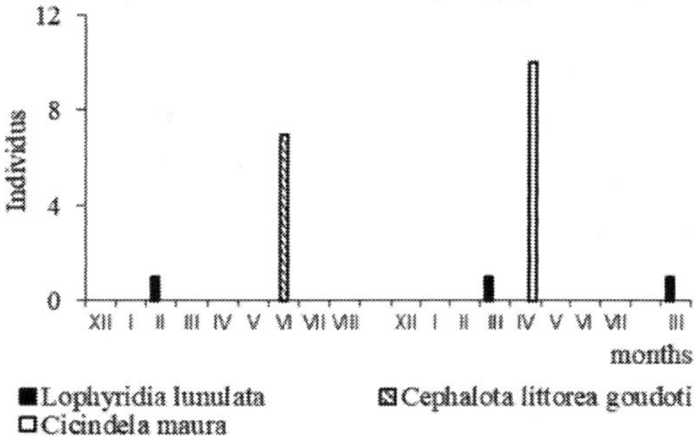

Fig. 2. Presence of Tiger beetles on the banks of daya EMDI depending on the month

Thus, *L. lunulata* is present during February and March, *C. maura* during April, and *C. littorea goudoti* during July. However, this succession was not noted by Chavanon [12], who observed the coexistence of tiger beetles on the banks of Moulouya River in perennial stations very close to ours.

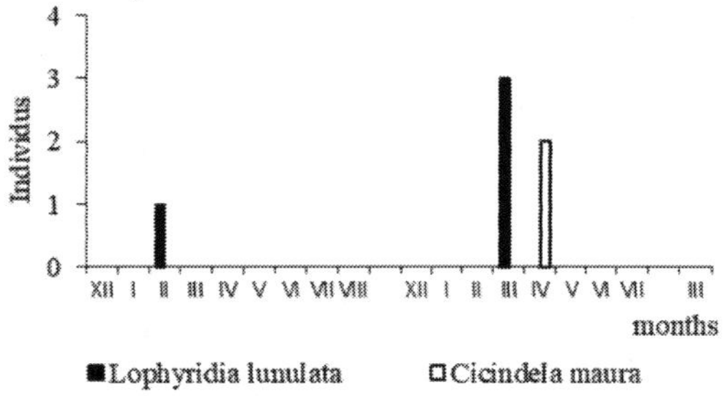

Fig. 3. Presence of Tiger beetles on the banks of daya EMDII depending on the month

A similar phenomenon occurs for species of genus *Pogonus* and *Pogonistes*. Indeed, it can be seen that despite an overlap in their period of presence on the banks of the EMDI daya, the populations reach their maximum expansion at different periods (Fig. 4).

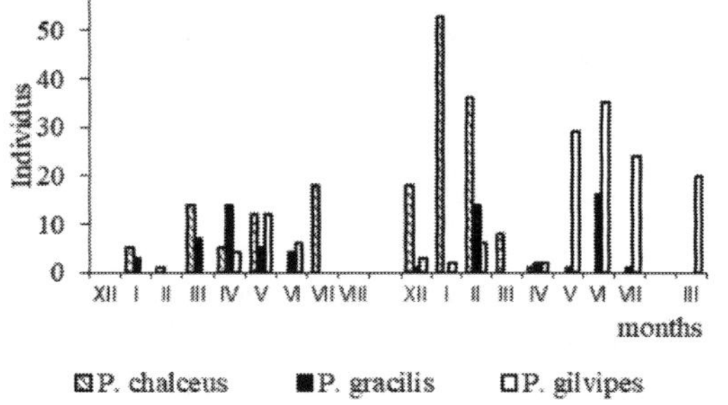

Fig. 4. Presence of some species on the banks of daya EMDI depending on the month

In the campaign I, *P. chalceus* dominates during July, while *P. gilvipes* peaks in May and *P. gracilis* during April.

In campaign II, a clear division of the proliferation period between P. chalceus dominates from December to March, and *P. gilvipes*, which proliferates from May onwards. *P. gracilis* is dominant in the middle of the cycle in both campaigns.

4 Conclusion

Banks of coastal dayas at the mouth of Moulouya are colonized by a diversified community, a mixture of species of terrestrial, aquatic and riparian origin. The latter dominate by their richness and abundance. They belong to *Trechidae* and *Scaritidae* families (*Caraboidea*) specialized in this type of environment [8, 9].

Constant species are not numerous and are halophilic species adapted to extreme conditions of these hypersaline environments.

Another type of adaptation characterizes these environments. It appears as a precise succession of tiger Beetles, as described in Spain [13], and a separate proliferation in time for Pogonus and Pogonistes. This is a particular phenomenon specific to temporary riparian environments. Indeed, these biotopes are very limited in space and time, and they force their populations to accelerate their life cycles so that species with the exact ecological requirements can succeed without competing.

References

1. W.D. Dudley, *The ecology of temporary waters*, Ed. Croom Helm : 183 (1987)
2. L. Bigot, G. Gautier, Originalité et intérêt écologique de la communauté ripicole et pélophile de surface. Bull. Mus. Hist. Nat. Marseille, **41** :13-30, (1981)
3. M. Antoine, *Coléoptères carabiques du Maroc : 1ère partie* Mém. Soc. Sc. Nat. Phys. Maroc, Zool., n. s., **1** : 1-177, (1955)
4. M. Antoine, *Coléoptères carabiques du Maroc : 2ème partie* Mém. Soc. Sc. Nat. Phys. Maroc, Zool., n. s., **3** : 1-137, (1957)
5. M. Antoine, *Coléoptères carabiques du Maroc : 3ème partie* Mém. Soc. Sc. Nat. Phys. Maroc, Zool., n. s., **6** : 1-150, (1959)
6. M. Antoine, *Coléoptères carabiques du Maroc : 4ème partie* Mém. Soc. Sc. Nat. Phys. Maroc, Zool., n. s., **8** : 1-180, (1961)
7. M. Antoine, *Coléoptères carabiques du Maroc : 5ème partie* Mém. Soc. Sc. Nat. Phys. Maroc, Zool., n. s., **9** : 1-153, (1962)
8. F. Rueda, C. Montes, Los Carabidos (Col. Carabidae) de las orillas de las lagunas salinas de la peninsula Iberica; Aspectos faunisticos. Actas **III** Congreso Iberico de Entomologia : 651-662 (1988)
9. J. Serrano, A.S. Ortiz, J. Gallian, Los Carabidae de lagunas y rios de la Submeseta Sur, España (Coleoptera, Adephaga). Bol. Asoc. Esp. Entom. Vol. **14** : 199-210 (1990)

https://doi.org/10.1051/e3sconf/202129803005

10. C. Favet, Contribution à l'étude écologique des populations ripicoles de Basse-Durance, Bull. Mus. Hist. Nat. Marseille, **44** : 47-61 (1984)
11. G. Gautier, Etude écologique des communautés de Tréchidés des bords d'étangs en Camargue Laguno-Marine : I- Le milieu et la faune (composition, structure et dynamique). Bull. Ecol., t. **10**, 3 : 181-202 (1979)
12. G. Chavanon, *Contribution à la connaissance des carabiques ripicoles de la Basse-Moulouya et de l'oued Za : Inventaire et distribution spatio-temporelle*, Thèse Doc. Etat, Univ. Mohamed Ier, Fac. Sci. Oujda : 159p. (1994a)
13. J. Vives, E. Vives, Coleópteros halóphilos de Los Monegros, Bol. Asoc. Esp. Entom. Vol **2** : 205-214 (1978)

 https://doi.org/10.1051/e3sconf/202129803006

Ecophysiology of the Mediterranean mussel Mytilus galloprovincialis L; effect of different microalgae diets and ration on broodstock conditioning

Hidaya Drissou[1], *Yassine* Ouagajjou[2], and *Adil* Aghzar[3,*]

[1]Faculty of Science, Abdelmalek Essaadi University, Tétouan 93002, Morocco
[2]Amsa Shellfish Station, National Institute of Fisheries Research (INRH), Amsa, Tétouan, 93022, Morocco
[3]Higher School of Technology, Sultan Moulay Slimane University, Khénifra 54000, Morocco

Abstract. Three diets based on three microalgae species (*Tetraselmi ssuecica*, *Isochrysis galbana,* and *Chaetoceros calcitrans*) and three feeding rations (1%, 2%, and 3%) were used to understand the ecophysiology of the Mediterranean mussel *Mytilus galloprovincialis* during Broodstock conditioning. Generally, mussels increase their consumption regarding food availability for all diets, and it is evident that the highest consumption was recorded with a feed ratio of 3%. During the study, *I. galbana* and *C. calcitrans* strains show the high consumption (1.58 µg. g^{-1}. h^{-1} ±0.00) and (1.58µg. g^{-1}. h^{-1} ±0.00) compared with *T. suecica* (1.542 µg. g^{-1}. h^{-1} ±0.023) recorded at 3% food ration of monospecific diets. Same results were recorded for ingestion but with great learning for *I. galbana* in triscpecific diet (0.70 µg. g^{-1}. h^{-1} ±0.052) compared to *C. calcitrans*(0.36 µg. g^{-1}. h^{-1} ±0.037) and *T. suecica* (0.25 µg. g^{-1}. h^{-1} ±0.126), respectively.The analysis of ANOVA has shown that both diet and food availability affect significantly the ingestion ($F = 30.9$, $df = 2$, $p < 0.001$ and $F = 4.4$, $df = 6$, $p < 0.001$) during mussel conditioning.

1 Introduction

The Mediterranean mussel Mytilus galloprovincialis distributed along the Moroccan coasts [1, 2] is an incredible gastronomic and economic interest species. The development of mussel farming is based on the availability of wild spat [3, 4]. Therefore, the availability of wild spat has played a crucial role in the mussel industry [5], and such availability is becoming a critical bottleneck because of the impact of scraping on natural beds [6]. This situation has given rise to different studies such as new spat collectors and rearing devices (e.g. [7]) and reproduction in captivity [8, 9].

The conditioning process is mainly based on the feeding management of broodstock and depends highly on the choice of the appropriate microalgae diets in terms of quality and food availability (ration) required for maturation in broodstock conditioning. Many studies on ecophysiology were performed to optimize feeding regimes and rations on conditioning and explain the effect of each physiological aspect [10, 11]. The present research aimed to evaluate the effect of multispecific microalgae diets and suitable food ration on physiological parameters like consumption and ingestion of the Mediterranean mussel broodstock.

2 Materials and Methods

2.1 Mussel sampling and broodstock conditioning

The biological material consisted of (28*10) adult specimens of the local mussel (weight = 23,41 ± 5,10 g; length =66,74 ± 5,86 mm) collected from natural populations from M'diq (23°47'26" N/15°44'26" M'diq, Morocco), then rinsed and placed in 20 L tanks and held in a closed water system. An average of 10 individuals per tank was used for each experiment under natural conditions of temperature (20°C), salinity (36 PSU), pH (8-8,3), and dissolved oxygen (6.4 ±0.5mg/L). Forty adults were randomly chosen in order to calculate their mean dry weight. Before carrying out feeding tests, all adults were held to fast for 48h in order to get rid of their feces.

2.2 Feeding process

Live microalgae used in this study consist of three species (*Isochrysis galbana*, *Chaetoceros calcitrans,* et *Tetraselmis suecica*). The choice of these three species is based on their high nutrient richness. Three diets, monospecific (M), bispecific (B), and tri-specific (T), were used during this experiment with different compositions. Three monospecific (M1:100% T.

*Corresponding author: a.aghzar@usms.ma

suecica; M2: 100% *I. galbana* and M3: 100% *C. calcitrans*), three bispecific (B1: 25% *T. suecica* + 75% *I. galbana*; B2: 25% *T. suecica* + 75% *C. calcitrans* and B3: 50% *I. galbana* + 50% *C. calcitrans*) and one Trispecific regimes (T: 25% *T. suecica*+ 75% *I. galbana* + 25% *C. calcitrans*). Furthermore, for each diet, we established seven feeding rations in triplicate as described by [12] (1%, 2%, 3% of dry mussel weight) in addition to controlling diet (0%: only natural seawater). The pseudofeces occurrence was taken calculated according to [13].

2.3 Statistical analyses

In order to estimate the effect of each treatment (diet and ration) on each calculated parameter, a two-way analysis of variance (ANOVA) was performed at a 95% level of confidence ($\alpha = 0.05$). When ANOVAs were significant, the post hoc Tukey multiple comparison tests were used to determine which treatments differed. All statistical tests were analyzed using the Rcmdr interface [14] implemented in R package version 3.1.2 [15].

3 Results

3.1 Consumption within diets and rations

Table 1 shows the feeding behavior of mussels in terms of consumption according to diets and rations. Generally, for all diets M, B and T, we observe that mussels increase their consumption regarding food availability, and the highest consumption was recorded with feed ration 3%. Within monospecific diets, mussels showed a small tendency towards *I. galbana*in terms of consumption (1.58 µg. g^{-1}. h^{-1}±0.00) and *C. calcitrans* (1.58 µg. g^{-1}. h^{-1}±0.00) strains compared to *T. suecica* (1.542 µg. g^{-1}. h^{-1}±0.023) at 3% food ration. For bispecific diets, and across all three diets tested, strains more consumed by mussel are *C. calcitrans* (1.18 µg. g^{-1}. h^{-1}±0.00) inDiet B2 and *I. galbana* (1.17µg. g^{-1}. h^{-1}±0.01)in Diet B1. However, *T. suecica* showed the lowest consumption (0.4µg. g^{-1}. h^{-1}±0.00) regarding the number of microalgae assimilated by mussels for all food rations and mixture diets. Whereas, the most important result obtained in this study is the high consumption (0.788µg. g^{-1}. h^{-1}±0.00) recorded in *I. galbana* in trispecific diet (T) at 3% food ratio (50% of total consumed biomass) and the other strains are equally consumed (25% of total consumed biomass).

Table 1. Consumption of microalgae (g dry weight algae) per dry weight of mussel M. galloprovincialis reared under different diets and food rations

Diet	Strain	Consumption (µg. g-1. h-1) ±SD		
		1%	2%	3%
M1	T. suecica	0.527 (± 0.000)	0.681 (±0.018)	1.542 (±0.024)
M2	I. galbana	0.525 (± 0.000)	1.027 (± 0.022)	1.58 (±0.000)
M3	C. calcitrans	0.526 (± 0.000)	1.052(± 0.000)	1.58 (±0.000)
B1	T. suecica	0.132 (±0.000)	0.26 (±0.00)	0.4 (±0.000)
	I. galbana	0.39 (±0.00)	0.78 (±0.06.10-4)	1.17 (±0.001)
	Total	0.52 (±0.00)	1.04 (±0.009)	1.57 (±0.001)
B2	T. suecica	0.131 (±0.000)	0.26 (±0.00)	0.4 (±0.00)
	C. calcitrans	0.39 (±0.00)	0.78 (±0.00)	1.18 (±0.00)
	Total	0.52 (±0.00)	1.04 (±0.00)	1.58 (±0.00)
B3	I. galbana	0.26 (±0.00)	0.52 (±0.00)	0.75 (±0.009)
	C. calcitrans	0.26 (±0.00)	0.52 (±0.00)	0.78 (0.06.10-4)
	Total	0.52 (±0.00)	1.04 (±0.00)	1.53 (±0.009)
T	T. suecica	0.13 (±0.00)	0.263 (±0.6.10-3)	0.39 (±0.00)
	I. galbana	0.263 (±0.00)	0.525 (±0.00)	0.788 (±0.00)
	C. calcitrans	0.13 (±0.00)	0.263 (±0.009)	0.4 (±0.00)
	Total	0.527 (±0.00)	1.052 (±0.6.10-3)	1.572 (±0.000)
Seawater		0.137 (±0.052)		

3.2 Ingestion within diets and rations

In terms of ingestion, generally, the occurrence of pseudofeces is almost rare in all cases, which means that mussels ingested the number of consumed microalgae. The statistical analysis of the effect of both diets and food availability on ingestion has been carried out during this study (table. 2).

Table 2. Effect of diets and food availability on ingestion during broodstock conditioning. Df (Degree of freedom), SumSq (Sum of Square), Sq (Square), F (F value), *** (significant at P< 0.001).

	Df	SumSq	Mean sq	F	Pr(>F)
Ration	2	6.2 x 10-12	3.10 x 10-12	30.9***	2.78 x 10-11
Diets	6	6.46 x 10-12	1.08 x 10-12	4.4***	4.8 x 10-4
Type of diets	2	6.26 x 10-12	3.13 x 10-12	12.7***	1.07 x 10-5
Ration/diets	20	6.3x 10 -12	3.24 x 10-12	31.5***	2.2 x 10-16

The food availability has most significantly influenced the ingestion of microalgae during mussels conditioning than diets (F=30.9 and F=4.4, respectively). However, we observed that the type of diets (M, B, and T) have also greatly influenced the ingestion of microalgae (F=12.7). In contrast, ingestion of microalgae was highly affected by both diets and ration (F=31.5).

4 Discussion

Food diets composition and rations (1%, 2%, and 3%) have shown different physiological parameters in the mussel broodstock. Both consumption and ingestion

https://doi.org/10.1051/e3sconf/202129803006

were increased within all diets regarding food rations that the food availability can explain. The tendency towards *I. galbana* and *C. calcitrans* strains compared to *T. suecica* can be explained by factors such as cell size, digestibility, and biochemical composition. Such factors determine the nutritive quality of the microalgae and their utility as food for bivalves [16, 17, 18]. Similar results were found in other bivalve species such as *Crassostrea virginica* [19], *Pectenmaximus* [20], and *Ventricosus circularis* [21].

For example, [22] demonstrated that the combination diet (*I. galbana* + *C. calcitrans*) at 22 °C was the best diet to give a maximal reproductive performance that fits our study well. This is explained by the fact that *I. galbana* had a higher percentage of lipids, but the combination with *C. calcitrans* was better than *I. galbana* alone due to the high percentage of carbohydrates accumulated in *C. calcitrans* cells. The microalgae *I.galbana* has reported an excellent food either in monospecific or multispecific diets [22]. Studies have shown that not all microalgae species are readily ingested by bivalve [23, 24]; the fact that particular microalgae species are consumed d[oes not necessarily mean they are ingested.

5 Conclusion

Despite the general recognition that a mixed microalgal diet gives better nutritional values for the conditioned bivalves, it is limited by the need for selection of algal strain used with suitable particle size, digestibility, and nutritive composition to get the total expected benefit of the mixed diets for bivalve hatcheries. Results and observations of the present study indicated that *I. galbana* and *C. calcitrans* had competitive conditioning performance, and the mixture of both had the highest conditioning potentiality as a diet for *M. galloprovincialis* broodstock conditioning. Therefore, *T. suecica* can be recommended for broodstock conditioning in the hatchery. *I. galbana* and *C. calcitrans* are recommended in bispecific diets, and *I. galbana* is more recommended in triscpecific diets. While, regarding food availability, more research is required to reveal the optimal ratio for mussels broodstock during the conditioning process.

References

1. H. Jaziri , T. Benazzou C. R. Biologies **325**: 1175-1183 (2002)
2. Y. Ouagajjou, P. Presa. Estuar. Coast. Shelf Sci., **152**: 1-10 (2015)
3. L.S Incze, R.A Lutz. Mussel culture: an east coast perspective. Pp. 99-140. In: R.A Lutz (ed.). Mussel Culture and Harvest: A North American Perspective. Elsevier Sci. Publ. Co., Amesterdam (1980)
4. R.W. Hickman, Mussel cultivation. In: The mussel Mytilus: ecology, physiology, genetics and culture [ed. by Gosling, E]. Amsterdam, The Netherlands: Elsevier Science Publishers BV.465-510 (1992)
5. J. Caceres-Martinez, A. Figueras, J Shellfish Res. **17**:153–157 (1998)
6. M. Holmer, K. Black, C.M. Duarte, N. Marba, I. Karakassis, Aquaculture in the Ecosystem. Netherlands: Springer (2008)
7. A. Aghzar, M. Talbaoui, M.H. Benajiba, P. Presa, Mar. Freshw. Behav. Physiol. **45**: 51–61 (2012)
8. P. Kamermans, T. Galley, P. Boudry, J. Fuentes, H. McCombie, F. Batista, A. Blanco, L. Dominguez, F. Cornette, Elsevier, New York, pp 339–373 (2013)
9. A. Aghzar, M. Miñambres, P. Alvarez, P. Presa, Thalassas **29**: 9–16 (2013)
10. J.I.P. Iglesias, M.B. Urrutia, E. Navarro, I. Ibarrola, J. Exp. Mar. Biol. Ecol. **219**: 71–86 (1998)
11. I. Martínez-Pita, C. Sánchez-Lazo, F.J. García, Aquacult. Nutr., **22** 202-216 (2016)
12. S.D. Utting, P.F. Millican, Aquaculture, **155** : 45–54 (1997)
13. R. Beiras, A. Pérez-Camacho, Mar. Biol. **120**:427–435 (1994)
14. J. Fox, Bouchet-Valat, M. Rcmdr. R Commander. R package version 2.3-1 (2016)
15. R DevelopmentCore Team. R: RFoundation for Statistical Computing, Vienna, Austria 2008, ISBN 3-900051-07-0, URL (2008)
16. G. Owen, Adv Comp Physiol Biochem **5** :1-35 (1974)
17. J.W Ewart, C.E Epifanio, Aquaculture **22**:297–300 (1981)
18. M. Albentosa, A. Perez-Camacho, U. Labarta, R. Beiras, M.J. Fernández-Reiriz Mar Ecol Prog Ser **97**: 261–269 (1993)
19. J. Babinchak, R. Ukeles Mar. Biol. **51**: 69-76 (1979)
20. M. Le Pennec, C. Rangel-Davalos, Aquaculture, **47** : 39 – 51 (1985)
21. M.C. Lora-Vilchis, A.N. Maeda-Martinez, Aquac. Res. **28** :905 – 910 (1997)
22. D. Matias, S. Joaquim, A.M. Matias, A. Leitão Invertebr Reprod Dev, **60**:49–58 (2016)
23. J. Babinchak, R. Ukeles, Mar. Biol., **51**: 69-76 (1979)
24. M.C. Lora-Vilchis, A.N. Maeda-Martinez, Aquac. Res. **28** : 905 – 910 (1997)

https://doi.org/10.1051/e3sconf/202129803007

Feeding and growth coupling during different development larvae phases of the Mediterranean mussel Mytilus galloprovincialis L. from Amsa Bay

Mohamed Ali El Moussaoui[1], *Yassine* Ouagajjou[2,*], *Adil* Aghzar[3], *El Mustafa* Ait Chattou[2], *Younes* Saoud[1], and *Hassan* Nhhala[4]

[1]Faculty of Science, Abdelmalek Essaadi University, Tétouan 93002, Morocco
[2]National Institute of Fisheries Research (INRH), Amsa 93022, Morocco
[3]Higher School of Technology, Sultan Moulay Slimane University, Khénifra 54000, Morocco
[4]Faculty of science and techniques, University of Abdelmalek Essaadi, Tanger 90060, Morocco

Abstract. Spat supply for mussel farming in Morocco is essentially based on seed capture and collection from natural beds. However, natural irregularity recruitment in exploited areas and encountered difficulties during spat collection operations have led to develop and improve reproduction techniques under controlled conditions in hatcheries. The present study focused on the larvae phase aiming at assessing growth rate varies between different development stages. Trocophore larva stage had a size of 63.25 μm and reached the pedivilegere stage with a size of 272.56 μm in 18 culture days. Length growth was more significant than width growth for both H-batch and T-batch. For instance, larvae grew in terms of length quickly in H-batch than T-batch (11.10± 1.30 and 9.67 ± 1.17 μm day^{-1}), and the same pattern was observed for T-batch larvae. Inversely, the total biomass consumed was higher during T-batch larvae rearing than H-batch ones. In contrast, the effect of feeding on growth was observed more significantly for H-batch larvae ($t=7.55$, $df=23$, $p<0.001$) than T-batch ones ($t=4.13$, $df=23$, $p<0.001$). However, the specific feeding effect was revealed higher for *I. galbana* for both batches rather than *C. calcitrans* and *T. suecica*. However, the larvae growth reduction is linked to the beginning of tissue synthesis.

1 Introduction

For decades, shellfish farming is mainly based on the availability of wild spat [1, 2]. The question of sufficiency in terms of spat supply for shellfish industry [3] linked to ecological problems [4] has emphasized studies related to hatchery spat production [5]. Optimization in bivalve hatchery productivity requires, among others, assessment of multi-species algae as feeding diets to enhance growth and survival during the larval phase.

Many studies have been carried out on bivalves to understand and improve the spat production process in this context. Such studies have successfully optimized conditions for the growth and survival of bivalve larvae in hatcheries [6, 7], while other studies have been based on physiological parameters [8]. Microalgae are the main source of nutrition in bivalves [9]. However, information on algae composition and larval food requirements for a high growth rate are limited [10]. Microalgae culture is so influential because it is the only feed source for bivalve larvae [11]. A large percentage of shellfish production cost is expended on feed [12, 13]; it represented almost 30% of the production cost [14]. Some commonly used micro-algae species as mussel larvae feed include *Isochrysis galbana, Isochrysis galbana Tahiti strain, Chaetoceros calcitrans, C. muelleri, C. gracilis, Tetraselmis suecica, Skeletonemacostatum, Rhodomonas lentil, Pavlova lutheri, and P salina* [15, 16].

Pelagic larvae require particular conditions for their development and metamorphosis to post-larvae (benthic larvae); this process requires an adequate food supply (type and concentration of phytoplankton), suitable environmental conditions, notably temperature [17] and physiological factors [18]. This research aims to summarize and review the latest literature on the development, feeding, metabolism, and growth of mussel larvae, emphasizing their physiological responses in their different stages until the settlement phase. Moreover, it highlights knowledge on the nutritional needs of mussels' larvae and improves their survival.

2 Materials and Methods

2.1 Spawning and fertilization

Spawning was induced triggered by a thermal technique where mussels broodstock were submitted alternatively to two different temperature seawater conditions, first cold (14 ± 1.5 °C) then hot (26 ± 1.0 °C). After repeating

*Corresponding author: ouagajjou@inrh.ma

https://doi.org/10.1051/e3sconf/202129803007

this treatment twice, males and females of mussel broods started laying their gametes; then, each spawned individual was placed separately in a single small container leaving them to continue releasing their gametes. Fertilization was performed based on a ratio of about 50 to 100 sperm per oocyte.

2.2 Larvae rearing

Freshly hatched larvae were reared in 0.2 µm filtered seawater under controlled conditions of temperature (22 ± 0.8 °C), pH (7.8 ± 0.15), and salinity (36.6 PSU). Larval rearing seawater was changed each day (48h), and larvae were sieved for sizes screening and daily growth rate estimates. Fifteen larvae were randomly sampled in each 48h renewal rearing seawater culture for biometry measurement (length and width).

2.3 Larvae feeding

Isochrysis galbana, Tetraselmis suecica. And *Chaetoceros calcitrans*) They were used as the three microalgae strains to feed reared larvae in this study. First, during the early larvae stages, only *I. galbana* and *C. calcitrans* were supplied as food based on 1,300 cells/larvae and 700 cells/larvae) respectively. Later, since culture day 8, *T. Suecica* was introduced in the feeding regime with 750 Cells/larvae.

2.3 Statistical analysis

Statistical analysis was performed using the Rcmdr interface [19] implemented in R package version 3.0.4 [20]. The test student was handled to reveal the effect of microalgae consumed biomass on larvae growth rate.

3 Results

Table 1 shows the length and width gain of larvae and their consumption variation in terms of micro-algae according to two slices head batch (H-batch) and Tail batch (T-batch). Generally, the H-batch larvae grow quickly (11.10± 1.30 µm day^{-1}) compared to the T-batch (9.67 ± 1.17 µm day^{-1}). Inversely, in terms of consumed biomass of microalgae, the latter showed a high consumption (4708.21 ng. $Larvae^{-1}$) compared to the H-batch slice (4053.81 ng. $Larvae^{-1}$) (Table 1). This can be explained by the fact that larvae use their consumed nutrients more in their tissues synthesis than in their growth. In terms of specific consumption, *T. suecica* is more consumed in biomass in both H-batch and T-batch with values of 2431.85 ng $Larvae^{-1}$ and 2970.43 ng $Larvae^{-1}$ followed by *I. galbana* with 1226.47 ng $Larvae^{-1}$ and 1290.75 ng $Larvae^{-1}$. respectively. On the other hand, *I. galbana* showed a significant effect of micro-algae consumption on the growth rate of mussel larvae during different development phases and both slices H-batch (T=9.21; df=23; p-value =3.55 x 10^{-9}) and T-batch (T=6.24; df=23; p-value =2.32 x 10^{-6}) (Table 2).

Table 1. Length and width variation for head batch (H-batch) and Tail batch (T-batch) regard consumed biomass of microalgae. LG (length gain), GR (growth rate).

		H-batch		T-batch	
		Length	Width	Length	Width
Larvae growth	LG (µm)	210.45± 3.29	180.45± 2.52	195.25± 15.05	162.5± 10.38
	GR (µm/day)	11.10± 1.30	9.95 ± 2.31	9.67 ± 1.17	6.67± 2.14
Larvae consumption (ng/larvae)	I. galbana	1226.47		1290.75	
	C. calcitrans	395.49		447.03	
	T. suecica	2431.85		2970.43	
	Total	4053.81		4708.21	

Table 2. Statistical analysis (test t) of the effect of microalgae consumption on the growth rate of mussel larvae during different development phases. T (test student), df (degree of freedom), *** significant at p< 0.001

t-test (student)	t	df	p-value
Head batch larvae			
I.galbana/growth rate	9.21	23	3.55 x 10^{-9}
C.calcitrans/growth rate	7.00	23	3.90 x 10^{-7}
T.suecica/growth rate	7.80	14	1.84 x 10^{-6}
Total consumption/growth rate	7.55	23	1.14 x 10^{-7}
Tail batch larvae			
I.galbana/ growth rate	6.24	23	2.32 x 10^{-6}
C.calcitrans/ growth rate	3.82	23	8.80 x 10^{-4}
T.suecica /growth rate	3.69	14	2.44 x 10^{-3}
Total consumption/growth rate	4.13	23	4.06 x 10^{-4}

4 Discussion

In this study, it has been noted that the consumption rate varied between the different larval development stages of larval development. It was equal to 84% in the D larval stage; then, it decreased to 68% in the viligere stage before increasing to 84 in the pediveligere stage. It has also been noted that consumption in the first stages depends on two species *I. galbana* and *C. calcitrans.* The consumption of *T. suecica* was small due to the contribution of the other two species. However, when larvae increase in size, particularly since the veliger stage, they begin to consume more *T. suecica* than *I. galbana* while their *C. Calcitrans* remains almost the same. These results indicated that the consumption rate is related to the morphological character of the alga itself, such as size and shape, which could make it challenging to be ingested by small size larvae spatially during early larvae development [18].

During the veliger phase, the growth has been increased from 10 to 20µm/day after the introduction of *T. suecica.* However, the growth decreased during the late larvae development phase, although the consumption rate continued to increase, specifically for *C. calcitrans* and *T.*

suiseca (97 and 80%, respectively). Moreover, the last days of larvae development, particularly in the pedivileger stage, are characterized by larval rapid morphological and physiological changes likely influenced by their growth [21]. Cells' size and morphology of used microalgae can determine feeding factors that influence pediveliger larvae. *I. galbana* and *C. calcitrans* are smaller in size than *T. suecica. H*owever, the sizes of these three algae are within the acceptable range of feeding for bivalve larvae, with an anteroposterior size higher than 210 μm [22, 13]. The higher biomass consumption observed for T-batch larvae could be clearly explained by the high heterogeneity of larvae batch (eyespot and pediveliger). During this late phase, the mainstream larvae were using nutrients for tissue and organ synthesis rather than growth.

Many authors reported that combining different microalgae species as a food promotes better larval growth and survival than a mono-species diet in some bivalve species [11, 15, 23]. Moreover, [14] have reported that *T. Tetrahele* as mono-regime for *P. viridis* larvae nutrition cannot meet their needs [13]. [24] have found that the use of *T. Suecica* gave a low growth rate in *P. generosa*. According to [25], feeding based on only one type of microalgae can affect the growth rate of larvae, while mixed diets feeding, composed at least by one type of flagellate and one type of diatoms, has been shown to produce optimal growth and development of bivalve larvae [25, 26, 27].

5 Conclusion

In conclusion, microalgae feeding plays a crucial role in mussels' production; it is considered the only nutrient source in the larvae culture of bivalves. As showed in this study, and according to scientific literature, growth performance is high when a mixed diet is given instead of a single source of feed. The increase in consumption rate reflected physiological need increases; however, when their size exceeds 150μm, larvae begin to prefer larger-sized microalgae with more nutritional components such as *T. suecica* and *C. calcitrans*. The growth rate remains constant in the pedivelligere stage characterized by morphological change, mainly foot appearance and shifting habits from pelagic to benthic life. In this phase, mussels' larvae prepare their fixation phase required for their metamorphosis process.

References

1. L.S Incze, R.A Lutz. Mussel culture: an east coast perspective. Pp. 99-140. In: R.A Lutz (ed.). Mussel Culture and Harvest: A North American Perspective. Elsevier Sci. Publ. Co., Amesterdam (1980)
2. R.W. Hickman, Mussel cultivation. In: The mussel Mytilus: ecology, physiology, geneticsand culture [ed. by Gosling, E]. Amsterdam, TheNetherlands: Elsevier Science PublishersBV.465-510 (1992)
3. J. Caceres-Martinez, A. Figueras, J Shellfish Res. **17**:153–157 (1998)
4. M. Holmer, K. Black, C.M. Duarte, N. Marba, I.Karakassis, Aquaculture in theEcosystem.Netherlands: Springer (2008)
5. A. Aghzar, M. Miñambres, P. Alvarez, P. Presa, Thalassas **29**: 9–16 (2013)
6. da Costa F., Cerviño-Otero A., Iglesias Ó., Cruz A., Guévélou E., Aquaculture International **28**:1675-1708 (2020)
7. Y. Ouagajjou, T. El Aloua, M. El Moussaoui, El M. AitChattou, A. Aghzar, Y. Saoud, AACL Bioflux, **13**: 3664-3673 (2020).
8. D. Matias, S. Joaquim, A.M. Matias, A. LeitãoInvertebrReprodDev, **60**:49–58 (2016)
9. Knauer J, Southgate PC. Rev Fish Sci 7:241–280 (1999)
10. Rico-Villa B, Le Coz JR, Mingant C, Robert R. Aquaculture **256**: 377–388 (2006)
11. I. Laing, Aquaculture 65,243–249 (1987)
12. P. Coutteau, FAO fish technical paper 361. FAO, Rome, pp 7–48 (1996)
13. M.M. Helm, N. Bourne, A practical manual. FAO fish technical paper 471. FAO, Rome, pp 59–83 (2004)
14. S. Rivero-Rodriguez, A.R. Beaumont, M.C. Lora-Vilchis, Aquaculture **263**, 199–210 (2007)
15. P.C. Southgate, A.C. Beer, P.F. Duncan, R. Tamburri, Aquaculture **162**: 247–257 (1998)
16. Ehteshami F, Romano N, RamezaniFard E, Hoseinzadeh SH (2016). AquacultNutr 23:671–680
17. Madrones-Ladja JA, dela Peña MR, Parami NP. Aquaculture **206**:313–321 (2002)
18. Webb KL, Chu FLE-special publication no. 2, Louisiana State University Division of ContinuingEducation, Lewes/Rehoboth Beach, Delaware USA, pp 272–291(1982)
19. J. Fox, Bouchet-Valat, M. Rcmdr. R Commander. R package version 2.3-1 (2016)
20. R DevelopmentCore Team. R: RFoundation for Statistical Computing, Vienna, Austria (2008)
21. J.r H. Maquirang, F L. Pedroso, M J Apines-Amar, L.A G. Piñosa, D C. Rendaje, J F. Cadangin, Fedelia F C. Merol · Carlos C. Baylon. Aquaculture, (2019)
22. Utting SD, Spencer BE. Laboratory leaflet number 68, Ministry of Agriculture, Fisheries and Food, Directorate of Fisheries Research (1991)
23. Aarab L, Pérez-Camacho A, del Pino V-Toledo M, de ViçoseGC,Fernández-P H, Molina L. AquacultInt21:1065–1076 (2013)
24. Liu W, Pearce CM, McKinley RS, Forster IP. Aquaculture452:326–341 (2016)
25. Galley, T.H., Batista, F.M., Braithwaite, R., King, J., Beaumont, A.R.,. Aquaculture International **18**, 315–325 (2009)
26. Laing, I., Ministry of Agriculture, Fisheries and Food Directorate of Fisheries Research, Lowestoft. (1991)
27. Martinez-Fernandez, E., Southgate, P.C., Aquaculture **263**, 220–226 (2007)

Micropropagation of zygotic embryos from genetically diverse almond seedling orchards in eastern Morocco

Souhayla Kodad[1,*], *Reda* Melhaoui[1], *Mohamed* Addi[1], *Christophe* Hano[2], *Nargis* Sahib[1], *Ahmed* Elamrani[1], *Malika* Abid[1], and *Aatika* Mihamou[1]

[1]Laboratoire d'Amélioration des Productions Agricoles, Biotechnologie & Environnement, (LAPABE), Faculté des sciences, Université Mohammed Premier, Oujda, Morocco
[2]Laboratoire de Biologie des Ligneux et des Grandes Cultures (LBLGC), INRAE USC1328, Université d'Orléans, 21 rue de Loigny la Bataille, F-28000 Chartres, France

Abstract. Numerous studies on perennial crops have shown that almond has a low tolerance to climate change, which led to detailed study for its adaptation by focusing on selecting heat-tolerant almond varieties. The long period of perennial agriculture creates particular challenges in a changing climate. The risk complicates the choice of a variety that the best variety for the current climate may be poorly suited for future climates. Hence the interest of having a large variability of tolerant genotypes. In the eastern Mediterranean region of Morocco, seedling almond trees are largely grown, which created an important gene pool, both for breeding programs and the selection of superior genotypes adapted to difficult pedo-climatic conditions. For the conservation of crop biodiversity in these almond groves, micropropagation is considered a feasible technique for producing and regenerating superior planting materials. Zygotic embryos of several local ecotypes of the almond native population known as "Beldi" were cultivated on Murashige and Skoog (MS) medium supplemented with different concentrations and combinations of auxins and cytokinins. The results showed multiple shoot initiation from zygotic embryos on MS medium containing 1mg/L of 6-Benzyl-aminopurine (BAP) combined with 0.5 mg/L of Indole butyric acid (IBA). A high rooting rate was obtained on a half-strength MS medium supplemented with 1mg/L of IBA.

1 Introduction

In Morocco, the almond trees are grown in several regions from North to South, under different environmental conditions, mainly in rain-fed areas with poor soils that receive little attention from farmers. The resulting variability of environment and climate has evolved into a wide variety of almond genotypes in each production region, as about 50% of the almond trees grown in Morocco are derived from seedlings [1]. Traditional almond cultivation used open-pollinated seedlings [2], which, with self-incompatibility, produced very high heterozygosity in this species [3]. This significant variability has provided a functional genetic pool for the evolution of almonds, allowing each growing region to select almond cultivars well adapted to the region [4]. The eastern region of Morocco was recently supported by the Belgian development agency "BTC" through the project "PROFAO *" to plant 6000 ha of almond trees and improve the post-harvest value chain, including the evaluation of almonds and derived products. Several studies have been undertaken to evaluate the genetic diversity and quality of nuts in the Moroccan almond population using morphological and phenological characterization [5,6]. However, studies on the *in vitro* multiplication of these almonds have, until now, never been done before.

Climate has a significant impact on temperate fruit and nuts, mainly in vulnerable regions like the Mediterranean, since the lack of chilling may directly affect perennial crops' yield. The currently focused studies on the selection of late-flowering cultivars and the temperature increase associated with climate change may cause considerable damages in the future, such as reducing yield performance and quality of almonds since several studies have revealed the chill decline caused by climate change [7,8]. Phenology is a significant determinant of the final almond yield [9] and plays a critical role in the processes involved. Hence, the importance of maintaining variability in almond crops in each production area seems to be a helpful tool for an adaptation strategy to future climate conditions, allowing sustainable production in variable weather conditions. Moreover, traditional almond plantations in eastern Morocco show a better richness in tocopherols than the other four introduced varieties cultivated in the same region (Fournat, Marcona, and the couple Ferragnes/Ferraduel) [10].

By in vitro culture, this study aimed to protect the variability of traditional almond plantations in eastern Morocco and will lead to a detailed study to preserve this

* Corresponding author: souhaila.kodad@gmail.com

exciting ecotype by proving its genetic variability for its rescue and preservation.

2 Materiels and methods

2.1 Plant material and explant preparation

In this study, zygotic embryos of several local genotypes were collected from old trees of almond growing in four areas in the North-East region of Morocco. These explants were surface-sterilized by immersion in a 40% (v/v) commercial bleach solution (NaOCI) for 20 min after pre-sterilization processes, which included washing with tap water for 3-5 min and then rinsing with 70% (w/v) ethanol for 30 s. Later, the mature seeds were rinsed with sterilized water five times for 5 min each.

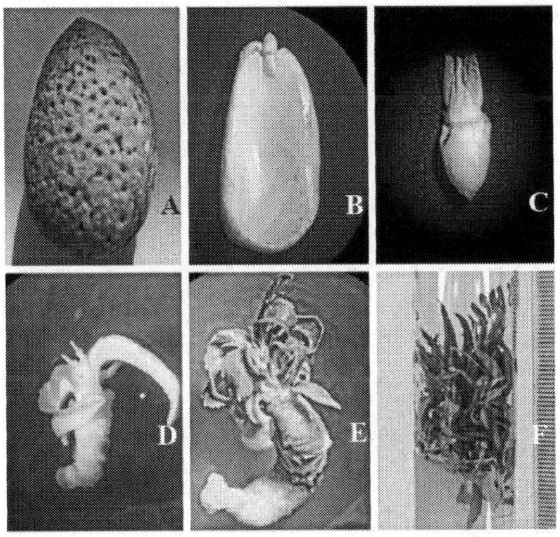

Fig. 1. Culture of zygotic embryos: (A) Beldi almond with shell (B) Half almond (C) Isolated zygotic embryo culture on MS medium + BAP 1mg/L (D) zygotic embryo after one week in the dark (E) zygotic embryo after been exposed to the light, (F) development of a cluster of shoots after 2 weeks.

2.2 Establishment of zygotic embryos in vitro culture

After sterilization, the zygotic embryos were isolated from the mature seeds before being inoculated into the Murashige and Skoog (MS) basal medium [11] supplemented with 1mg/L of 6-Benzyl-aminopurine (BAP) or Kinetine (KIN) associated or not with 0.5mg/L Indole butyric acid (IBA) (Fig. 1b, c). One week after culture in the dark (Fig. 1d), the germinated almond embryos were exposed to the light, and after four weeks of culture, they produced actively growing shoots (Fig. 1e, f). In the first stage, for *in vitro* proliferation of axenic germinated explants with at least an apical shoot, the explants were aseptically cut and cultured in a hormone-free MS medium. The newly formed shoots were separated and used for the induction of *in vitro* rooting.

2.3 Rooting and acclimatization

For rooting experiments, two different media (Half-strength MS and Heller) containing different concentrations of IBA (0, 0.5, and 1 mg / L) were tested (Table 2). The media were supplemented with 3% sucrose (w/v) and solidified with agar (0,7%, w/v, Agar-Agar, sigma) adjusted to pH 5.7 before autoclaving (120°C for 20 min). Plant growth regulators were added to the medium before the adjustment of pH and sterilization. The cultures were maintained at 25 ± 2°C with a 16 h photoperiod (40 μmol m-2 s-1) provided by mercury fluorescent lamps. *In vitro* rooted shoots were washed in running water before being potted in a 1:1:1 mixture of perlite, sand, and soil. Plantlets were covered with plastic cups to maintain relative humidity for 3-4 weeks before transfer into the growth room. The growth room was illuminated by mercury fluorescent lamps (400 W). The plants were irrigated every 2-3 days with water and, after 38 days, were successfully adapted to the *in vivo* condition.

2.4 Statistical analysis

The experimental design used in this study was completely randomized. Analysis of Variance (ANOVA) analyzed all data, and Duncan's multiple range tests were performed to analyze shoot number and shoot length rates. Likelihood ratio $\chi 2$-test (G-test) was used to analyze rooting frequencies. Living sterile shoots and roots constituted the basis of the statistical analysis. For all comparisons, statistical significance was considered at P B 0.05.

3 Results and discussion

3.1 Effect of plant growth regulators on zygotic embryos in vitro culture

In the first experiment, several zygotic embryos (B1, B2, B3, and B4) isolated *in vitro* conditions were cultured on MS medium supplemented with BAP or KIN either alone or in combination with IBA or NAA. Germination rate, as well as the number and the length of shoots, were recorded (Table 1). The effect of type of cytokinin on the culture initiation of zygotic embryos isolated from mature seeds of almonds significantly affects the frequencies of embryos germination. However, either in BAP or KIN, high germination rates were obtained, ranging between 70.5 and 93.05% (Table 1). The presence of BAP has a higher effect than KIN according to the number and the length of new shoots formed since the number of shoots was between 3 and 4 in the presence of BAP. While with KIN, it was between 2 and 3, and for the length of shoots, it ranged from 2,5 to 3,1cm in the presence of BAP, while with KIN, it was 2 to 2,4cm. For most genotypes, the combination of 1mg/L of BAP or KIN with 0.5 mg/L IBA or NAA resulted from significant differences. Indeed, the best germination rates were obtained with B3 genotype on MS medium supplemented with 1mg/L BAP and 0.5 mg/L IBA (93.05%), while the maximum number of shoots was obtained in the presence of BAP at 1mg/L combined with IBA at 0.5mg/L (9 shoots) and a substantial length was also recorded at the same concentration (3.1cm). While in the presence of NAA associated with KIN, the highest germination rates (90.2%) and length (3.0cm) were obtained with the B4 genotype, and the maximum number of shoots (6.43) was recorded for the B1 genotype at the same concentrations.

Table 1. The germination rate, the number and length of shoots of isolated zygotic embryos according to the concentrations of different phytohormones.

Genotypes	Plant Growth Regulators (mg/L)				Germination rate (%)	Average shoot numbers	Average shoot Lengths (cm)
	BAP	KIN	NAA	IBA			
B1	1	0	0	0	84.07g	3.1o	2.2f
	1	0	0.5	0	83.1h	4.0i	2.5c
	1	0	0	0.5	91.02b	5.2f	3.1a
	0	1	0	0	78.5k	2.33p	2.4d
	0	1	0.5	0	79.29j	6.43b	1.5i
	0	1	0	0.5	89.2e	4.0i	2.0h
B2	1	0	0	0	80.04i	3.67j	3.1a
	1	0	0.5	0	77.5l	5.4d	2.4d
	1	0	0	0.5	85.28f	6.0c	1.5i
	0	1	0	0	79.2j	3.17n	2.0h
	0	1	0.5	0	81.03i	5.4d	2.3e
	0	1	0	0.5	90.17d	3.3l	2.5c
B3	1	0	0	0	80.03i	4.0i	2.3e
	1	0	0.5	0	90.12d	5.2f	2.5c
	1	0	0	0.5	93.05a	9.4a	3.1a
	0	1	0	0	70.5m	2.12q	2.4d
	0	1	0.5	0	79.39j	3.5k	1.5i
	0	1	0	0.5	84.05g	5.1g	2.3e
B4	1	0	0	0	79.34j	2.02s	2.5c
	1	0	0.5	0	84.05g	4.6h	2.3e
	1	0	0	0.5	90.1d	3.2m	2.5c
	0	1	0	0	80.07i	2.04r	2.1g
	0	1	0.5	0	90.2b	5.3e	3.0b
	0	1	0	0.5	86.1ef	6.8b	2.6c

Different lowercase letters above any two columns indicate that the means are statistically different at P≤0.05 according to the Duncan multiple tests. Data were taken two weeks after culture.

Previous studies have shown that cytokinins alone or auxins are essential for explant proliferation [12]. Our results agree with those obtained in other research where the best shoot growth and proliferation rate of the commercial cultivars of almond (Nonpareil and Texas) were recorded in MS medium containing 1.0 mg/L of BAP [13]. Similar results were reported by Tabachnik and Kester (1977) [14], showing that the use of BAP in the shoot multiplication was an absolute necessity and the reduction of BAP concentration inhibited new axillary shoot formation. In comparison, we added IBA at the concentration of 0.5 mg/L to the medium to obtain significantly higher results with a maximum number of shoots (9) and significant lengths (3 cm).

3.2 Rooting and acclimatization of zygotic embryos

For rooting experiments, two different media (Half-strength MS and Heller) containing different concentrations of IBA (0.5, 1, and 1.5 mg/L) were tested (Table 2).

Table 2. Rooting of the Beldi ecotype in various media and different concentrations of IBA.

Rooting media		Rooting rate%	Roots number	Roots length (cm)
Media	IBA (mg/L)			
Half-strength MS	0.5	21,12e	1.0d	1.0c
	1.0	62,43a	6.0a	2.3a
	1.5	43,71b	2.0b	1.7b
Heller	0.5	15,03f	1.0d	0.3b
	1.0	28c	2.0b	0.5d
	1.5	25,1d	1.5c	1.4b

Different lowercase letters above any two columns indicate that the means are statistically different at P≤0.05 according to the Duncan multiple tests. Data were taken six weeks after culture.

The half-strength MS with 1 mg/L IBA was found to be the best medium for *in vitro* rooting of the "Beldi" ecotype because it gives the highest rooting rate (62,43%) and produces the maximum number of roots (6), and it had significant lengths (2.3 cm) (Table 2).

The use of auxins for *in vitro* rooting has also been reported by other studies; in a previous study on GF677, the highest rooting percentage was found on a medium containing 1mg/L of IBA. It has also been reported that IBA induces better lateral rooting than indole acetic acid (AIA) [15], [16].

4 Conclusion

In vitro culture techniques could be an essential tool for the production and regeneration of superior planting materials with a high tolerance to climate-changing, and it could be a unique solution for the conservation of crop biodiversity in almond groves by eliminating various diseases and increasing the variability of tolerant genotypes to maintain a sustainable production in the current climate in the future ones.

In this study, the regeneration of different explants was significantly affected by the genotypes and the different plant growth regulators combinations and concentrations. The results showed multiple shoot initiation from embryos on MS medium containing 1mg/L BAP combined to 0,5 mg/L IBA and the best *in vitro* rooting of the "Beldi" ecotype was obtained in a half-strength MS medium supplemented with 1 mg/L IBA.

References

1 A. lansar. H.Azoulay,D. E. Kester, *II Int Sym P A.* 96–100 (1998)

2 K. Sorkheh, S. Kiani, A. Sofo. *Fd ch,* **212**, 58-64 (2016)

3 R. Socias I Company, A. J. Felipe, *J. Hortic. Sci.* **67**, 313–317 (1992)

4 T. M Gradziel,. *A B P.* Cabi.. (2017).

5 A. Mahhou, F. G. Dennis,*Hort Tech*, **2**, 4, 488-492 (1992)

6 A. Lansari, A. F. Iezzoni, D. E. Kester, *Euph*, **78**, 1, 27-41 (1994)

7 A. Rodríguez, D. Pérez-López, E. Sánchez, A. Centeno, I. Gómara, A. Dosio, *Nat. Hazards Earth Syst. Sci.* **19**, 1087–1103 (2019)

8 R. Darbyshire, E. W. R. Barlow, L. Webb, I. Goodwin in (2014) XXIX Int. Hortic. Congr. Hortic. Sustain. Lives, Livelihoods Landscapes **1130**, 11–18. (2014)

9 I. J. Lorite, J. M. Cabezas-luque, O. Arquero, C. Gabaldón-leal, C. Santos, A. Rodríguez *Agric. For. Met.* **294**, 108142 (2020)

10 R. Melhaoui, M. L. Fauconnier, M. Sindic, M. Addi, M. Abid, A. Mihamou, A. Elamrani, *N Symp App Bio Sc,* (2018)

11 T. Murashige, *Symp. Tissue Cult. Hortic. Purp.* **78**, 17–30 (1977)

12 F. Sadeghi, A. Yadollahi, M. J. Kermani, M. Eftekhari, *J. Genet. Eng. Biotechnol.* **13**, 19–23 (2015)

13 S. Gürel, Y. Gülşen, *Turk. J. Bot.*, **22**, 363–373 (1998)

14 L. Tabachnik, D. E. Kester (1977)

15 M. N. Nas, Y. BÖLEK, N. Sevgin, *Turk. J. Bot.* 1134–1144 (2013)

16 G. J. Klerk, W. Van Der Krieken, J. C. de Jong. Review the formation of adventitious roots: new concepts, new possibilities. *IVC & D Bio-Pl*, **35**, 3, 189-199 (1999)

https://doi.org/10.1051/e3sconf/202129803009

Micropaleontological study of the Moroccan Mediterranean western coastline foraminifera and their morphological abnormalities

Haitam Afilal[1, *], *Abdellah* Trankil[1], *Saida* Bouzid[1], and *Abdelhamid* Rossi[1]

[1]Faculty of Sciences and Techniques of Tangier, Abdelmalek Essaadi University, Tetouan, Morocco.

Abstract. In order to contribute to the identification of the current faunistic composition of the Moroccan Mediterranean western coastline foraminifera, sediment from 4 stations have been analyzed, along with some abiotic parameters (pH, dissolved oxygen, temperature, and salinity), In order to characterize their populations, different Ecological indices were calculated and compared in the various samples. A total of 3451 individuals belonging to 14 species were recognized (90,26% hyaline, 9,74% porcelaneous, and none agglutinated test was found), all assemblages were dominated mainly by *Elphidium crispum*, *Ammonia beccarii*, and subordinately *by Quinqueloculina sp1* and *Heterolepa dutemplei*. All sampling sites showed remarkable similarity in the Foraminiferal assemblages, and the Oued Nigro station is considered the wealthiest and densest site, whereas Oued Martil records the lowest values of those parameters. On the other hand, the Sample from M'diq presents the lowest diversity index even if it is the second in terms of density. The faunistic analysis had also allowed the observation of some morphological abnormalities only found in the Oued Nigro station, affecting some individuals' tests. A small-sized chamber mainly manifests the abnormal appearances in the spiral whorls and calcification defects in the calcareous wall.

1 Introduction

At the interface of oceanic, terrestrial, and atmospheric domains, coastal zones offer a unique ecological richness, where the foraminifera assemblages represent more than 50% of the total sediment biomass [1].

Although current foraminifera species have been widely studied in the Mediterranean Sea [2-3], their assemblages in the Moroccan coasts are still poorly known. In this context, this study aims to describe the faunistic composition of Foraminifera sheltering the Moroccan Mediterranean coastline fringe. To this end, sediment from 4 stations has been analyzed.

2 Objectives

Foraminiferal assemblages present a great scientific interest, given their great potential as biostratigraphical and environmental indicators.

Therefore, this study is of great scientific interest since it aims to draw up an inventory of the species sheltering the Mediterranean coastal areas and identify their morphological abnormalities.

3 Methodology

3.1 Study area and sampling sites

The study area is located between Oued Nigro and Oued Martil, extending along 25 km of coast. In April 2021, a total of 5 samples were collected from 4 stations along the coastline fringe. Sampling points locations were chosen based on their proximity to towns, ports, and discharges from continental emissaries.

One sample was collected from each station 10 meters from the coastal fringe during low tide conditions, except Oued Nigro's station, where an additional sample was taken from its mouth. (Fig. 1)

3.2 Field sampling

Samples were collected and stored in plastic bags, using a 75 mm diameter PVC core 10 cm deep. pH, dissolved oxygen, temperature, and salinity data of the surface water from each sampling point were measured using a multiparameter.

* Corresponding author: afilal.haitam@etu.uae.ac.ma

Fig. 1. Location of the sampling stations along the Moroccan Mediterranean Western coastline.

3.3.2 Individuals extraction and identification

Each sample was spread on a picking tray, and individuals were picked systematically from side to side along the rows. We were using a thin brush under a binocular loupe. All individuals were picked from the sample and placed in cells to identify based on previous studies on benthic foraminiferal assemblages [4-5].

4 Results and Discussion

4.1 Foraminifera inventory

Three thousand four hundred fifty-one individuals belonging to 14 species were recognized (90,26% hyaline, 9,74% porcelaneous, and none agglutinated test was found). All assemblages were largely dominated by *Elphidium crispum* (49,58%), *Ammonia beccarii* (26,95), and subordinately by *Quinqueloculina sp1* (9,34%) and *Heterolepa dutemplei* (7,01%). (Table 1, Fig .1)

Table 1. Distribution (in absolute numbers) of the foraminifera species of the Mediterranean Western coastline in 50 g of sediment (the table indicates the corresponding identification used in the inventory plate).

Species	ID	Station				
		Oued Negro		Kabila	M'diq	Martil
		ON1	ON2	K1	MD1	MA1
Elphidium crispum	A	238	881	94	490	8
Ammonia beccarii	B	210	349	131	236	4
Ammonia tepida	C	4	14	0	0	0
Heterolepa dutemplei	D	19	94	11	117	1
Lenticulina vortex	E	2	31	0	33	0
Gyroidina leavigata	F	6	81	5	21	0
Gen 1	G	2	33	0	0	0
Triloculina sp1	H	2	4	4	3	1
Quinqueloculina sp1	I	44	170	29	71	8
Cancris auriculus	J	0	3	0	0	0
lenticulina sp2	K	0	2	0	0	0
Quinqueloculina sp2	L	0	1	0	0	0
Cibicides	M	0	1	0	0	0
Globigerina bulloides	N	0	1	0	0	0
Total	-	27	1665	274	971	22

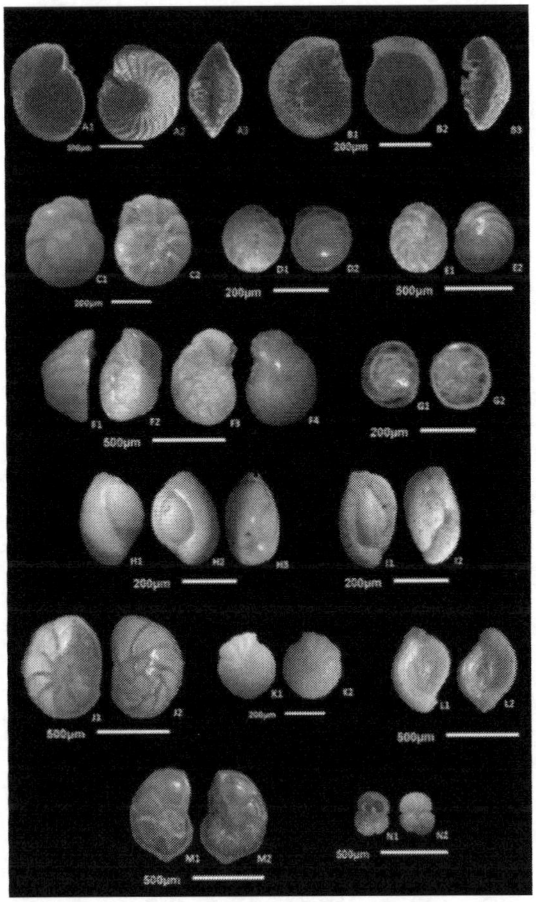

Fig.1. Photos of foraminifera species from the Mediterranean coastline fringe between Oued Nigro and Oued Martil.

 https://doi.org/10.1051/e3sconf/202129803009

4.2. Physico-chemical parameters and ecological indices

For each station, the temperature, dissolved oxygen, salinity, and pH data have been measured to define the environmental conditions during sample collection. (Table 2)

In order to characterize the faunistic composition, the total foraminiferal density and the specific richness (number of different species per sample) were calculated; as for the diversity, we used the Shannon–Wiener index, which links the number of species to the assemblage density. (Fig. 2)

Table 2. The physicochemical parameters of the shallow water at each sampling point.

Parameters	Station				
	Oued Negro		Kabila	M'diq	Martil
	ON1	ON2	KA1	MD1	MA1
Temperature c°	20	18,4	18,3	19,7	19,3
O2 dissolved mg/l	10,5	7,22	6,8	7,96	4,92
Salinity ppm	28,6	35,2	36,87	37,13	27,73
pH	8,25	8,4	8,37	8,33	7,97

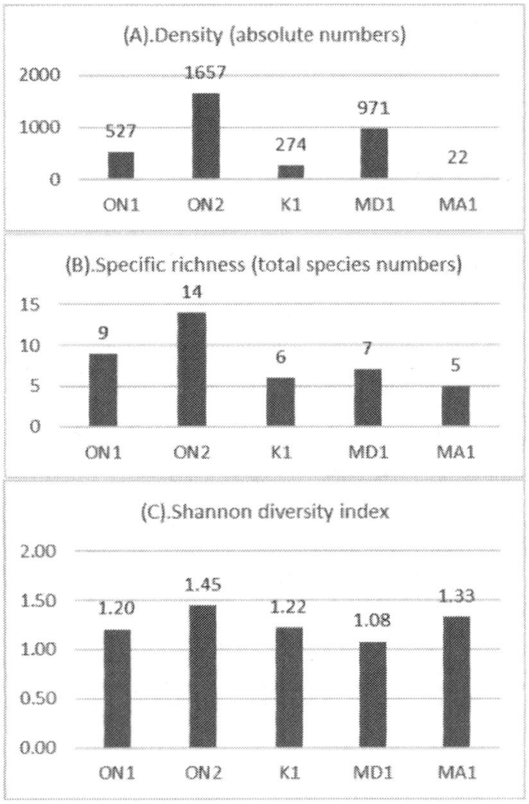

Fig. 2. (A) Total foraminiferal density, (B) the specific richness, (C) the Shannon–Wiener diversity index for the four sampling stations.

The samples analyses showed significant similarity in the foraminiferal assemblages among all stations—moreover, highly variable faunal densities vary between 22 to 1665 individuals.

Increasing in the specific richness (9 to 14) and density (527 to 1657) from Oued Nigro mouth towards the sea is observed, considered the richest and the densest site among all samples.

The very low density and richness are found at Oued Martil station with only five species and a total of 22 individuals,

On the other hand, the Sample from M'diq presents the lowest diversity index even if it is the second in terms of density.

The increase of individuals and species from upstream to downstream in the Oued Nigro sediment is due to the increasing salinity in the water since it represents a factor of influence on the distribution of the foraminifera assemblages [6].

Regarding the Oued Martil station, apart from its low density compared to other stations, most of the individuals were degraded and fractured, and probably none of them have been alive during the samples collection due to the great pollution load received in this region, according to many studies [7].

The physicochemical parameters collected shows a normal environmental condition, and therefore the observed changes in the foraminiferal associations may be due to other factors as hydrodynamic, granulometric, organic matter concentration

4.3 Morphological abnormalities

During the faunistic analysis, some abnormalities affecting some individuals caught the attention, especially that those abnormalities were seen only in the Oued negro station. The abnormal appearances observed had affected only the abundant species *Elphidium crispum* and *Ammonia beccarii*. Those abnormalities are mainly represented in a small size chamber in the spiral whorls and calcification defects in the calcareous wall. (Fig. 3)

4.3.1 Small size chambers

It results by the presence of one chamber or more with a smaller size than the average, thus breaking the regularity of the spiral (Fig. 3: a1, a2, b1, b2), showing that at the time when the individual was building or preparing to build this chamber a physiological disorder occurred. The small chamber does not always occupy the same rank in the spire. This can be explained by the fact that individuals in the same population do not all appear simultaneously and that, in the same generation, they do not necessarily have a synchro growth.

4.3.2 Calcification defects

The Calcification defect is manifested in the form of folding or crumpling in the calcareous wall, as if the wall had been «softened», then recalcified while keeping a «crumpled» aspect. (Fig. 3: b3, b4). This anomaly only concern *Ammonia beccarii*, and it does not appear to be due to a crash caused by a shock because no fracture was detected

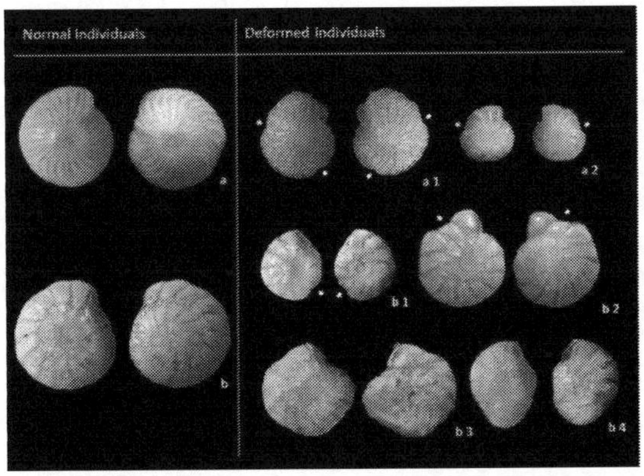

Fig. 3. Photos of foraminifera species with morphological abnormalities.

The small chamber of the individuals experiencing a slowdown in growth corresponds probably to the first chamber built after a disturbance in the conditions of the environment. The crumpling aspect may indicate transient acidification of the environment. Since the calcareous wall behaves like it was partially decalcified, collapsed, then recalcified, keeping a deformed aspect.

It appears that the Mediterranean western coastline fringe has a similar foraminiferal composition with a highly variable faunal density.

The morphological abnormalities detected in some individuals' tests show how sensitive these organisms are to environmental disturbances. All the abnormal appearances can be recorded thanks to the fossilized nature of tests in sediment after death.

5 Conclusion

The determined foraminiferal assemblages will give reliable indications on the environmental quality since the ecological characteristics of the different species are different enough to obtain reliable information about the environmental conditions, as shown in a wide range of papers on benthic foraminiferal ecology [8-9].

Apart from the faunistic analysis, this study allowed detecting a certain number of deformations of some individual's test. The number of individuals deformed is low since the samples were taken for a completely different purpose than morphological analysis. They were intended for the inventory of the microfauna. That is why the coring sampling method was chosen. Otherwise, a sampling by skimming the surface could have been more relevant for the morphological study. It should be noted that the detection of abnormalities is not always possible; a certain number of anomalies likely to affect first chambers can be masked by the embracing chambers of the subsequent whorls giving a regular aspect even that it is deformed from the inside.

Acknowledgments

I am pleased to address a special thanks to Pr. Rezqi Halima from the Mohammed Premier University of Oujda for her help in verifying the species identified.

References

1. A.J. Gooday, A.E. Rathburn. Temporal variability in living deep-sea Foraminifera: a review. Earth-Science Reviews 46, 187–212. (1999)
2. J.P. Debenay, E. Tsakiridis, R. Soulard, H. Grossel,. Factors determining the distribution of foraminiferal assemblages in Port Joinville Harbor (Ile d'Yeu, France): the influence of pollution. Mar. Micropaleontol. 43, 75–118 (2001)
3. R. Coccioni, F. Frontalini, A. Marsili, D. Mana, Benthic Foraminifera and trace element distribution: a case-study from the heavily polluted lagoon of Venice (Italy). Mar. Pollut. Bull. 59, 257–267 (2009)
4. A.R. Loeblich, H. Tappan, Foraminiferal genera and their classification. New York: Van Reinhold. (1987)
5. J.W. Murray, Ecology and distribution. In: BENTHOS'90, Tokai University Press, Sendai. (1991)
6. J.W. Murray, Syndepositional dissolution of calcareous Foraminifera in modern shallow water sediments. Marine micropaleontology 15, 117–121 (1989)
7. M. Blinda, Land-based pollution of the northwest coast of Morocco between Tangier and Tetouan: Characterization, Impact on the Environment and Solutions Proposal (2007)
8. J.W. Murray. Ecology and Applications of Benthic Foraminifera. Cambridge University Press, Cambridge, **426**. (2006)
9. F.J. Jorissen, C. Fontanier, E. Thomas. Chapter seven paleoceanographicalproxies based on deep-sea benthic foraminiferal assemblage characteristics. In:Hillaire-Marcel, C., De Vernal, A. (Eds.), Paleoceanography of the Late Cenozoic.Developments in Marine Geology, Elsevier,263–325 (2007)

Applying Geographic Information Systems (GIS) for surface condition indicators modeling of a flexible pavement

Mohammed Amine Mehdi[1,2,3,*], *Taoufik* Cherradi[1], *Said* El Karkouri[2], and *Ahmed* Qachar[3]

[1]Mohammadia School of Engineers, Civil Engineering and Construction laboratory, Rabat, Morocco.
[2]National Center for Road Studies and Research, Ministry of Equipment, Transport, Logistics and Water, Rabat, Morocco.
[3]Moroccan Road Directorate, Ministry of Equipment, Transport, Logistics and Water, Rabat, Morocco.

Abstract. In Morocco, as in all world countries, deteriorating road conditions, increasing traffic loads, and decreasing funds have presented a complex management challenge for the road maintenance and rehabilitation process. Hence the need to assess the condition of the pavement network, decide on maintenance strategies, set rehabilitation priorities and implement a maintenance management system. In this regard, Geographic Information System (GIS) is a powerful tool for managing and analyzing data referenced to a geographic location, especially in the field of road infrastructure where information on pavement sections stored in textual databases can be linked by location and attribute in geographical maps. This paper presents a case study, explores the ability of a GIS to visualize the different levels of surface degradation of flexible pavement through the analysis of GIS surface indicator matrices and the reduction of road databases containing the results of the environmental inspection carried out in 2018 on a 50 km section of the Moroccan national road number 06 from KP 0+080 to KP 0+130, applying the Moroccan method carried out by the Moroccan National Center for Road Studies and Research. These thematic maps can justify a budgetary request for the investment of public funds and help in maintenance decisions.

1 Introduction

The current and future needs for road data manipulation and geographic information applications are enormous. Indeed, road network management and its optimization are operations complicated to manage with the usual supports such as plans, maps, or photographs. The road management system is based on the phenomenon of the information system currently represented in geographic databases, also called spatial reference in forms that require a diverse process of integration adequate and thoughtful. The use of geographic information systems is characterized by the exchange of data and the circulation of spatial information between actors operating on a territory.

Fig. 1. Cracks

As part of our collaboration with the Moroccan National Center for Road Studies and Research [1], this article will help reduce the effort required in conventional methods to collect and analyze data periodically by reducing repetitive work. The present work analyses the surface deteriorations such as potholes, pullouts, and cracks (Fig. 1, 2, and 3).

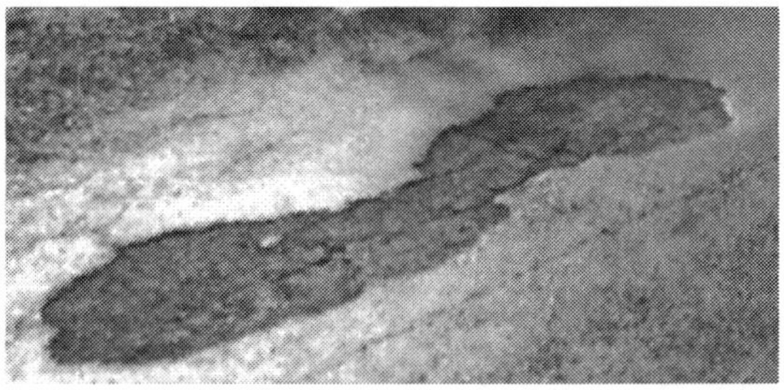

Fig. 2. Pullout

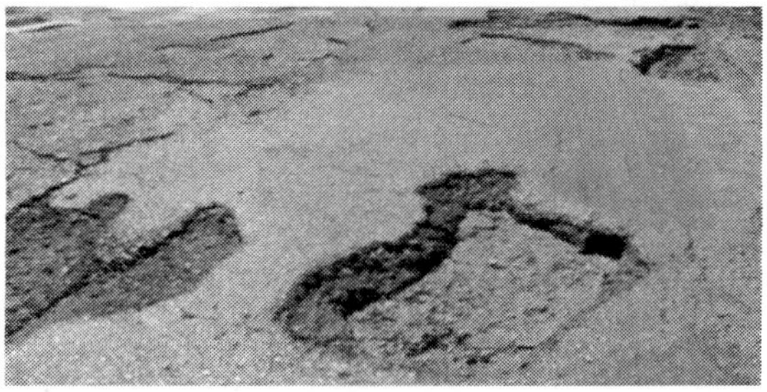

Fig. 3. Potholes

* Corresponding author: aminemehdi@research.emi.ac.ma

https://doi.org/10.1051/e3sconf/202129804001

The section under study is part of the Moroccan National Road number 06, from KP 0+080 to KP 0+130.

Given the geographical location, this road link has known since its birth a high traffic demand estimated at 3.7 Million AADT in 2017. The essential objectives of this article are:

1. Reduce the data of the environmental inspection carried out in 2018,
2. Represent the equi-qualities of this pavement according to the surface indicator,
3. Modeling the thematic map by of the surface indicator of this pavement by GIS.[2]

2 Surface condition indicator (SUI):

2.1 Moroccan method

Every two years based on a visual survey of pavement deteriorations. The visual survey concerns deterioration of the following types: cracks (longitudinal and transverse) [1-3], tears (feathering, peeling, combing, potholes), deformations (rutting, subsidence, etc.), and shoulder slopes. During the same survey, the wearing course nature, the roadway width, shoulders, the traffic class, the nature, and the date of the last intervention are also noted. The principle of the survey is based on the observation at regular intervals of 200 m of the existence or absence of each type of deterioration, indicating its severity (Fig. 4).

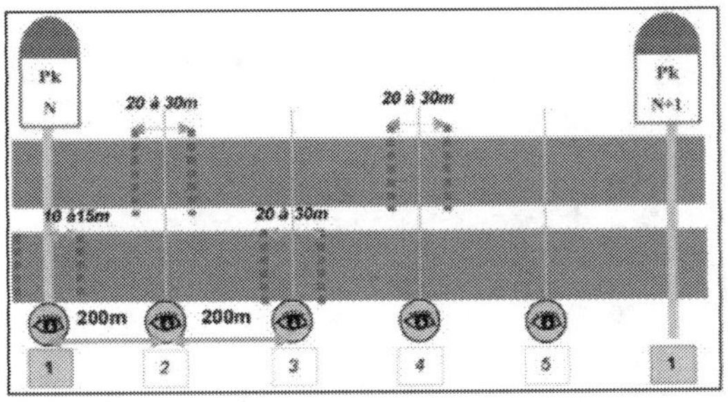

Fig 4. Visual Moroccan inspection method

The integration of the observations is done by adding up the scores given to each of the deteriorations (cracks, pullouts, and potholes) on each 1km section.

These indicators characterize 1 km of road and can have four values:

- A: Good
- B: Acceptable
- C: Poor
- D: Very bad

3 Materials and methods

3.1 Inspection

For each km of pavement, the surface distress condition [4] is classified into four classes for each distress according to the following decision grid, based on the cumulative scores over the km:

Table 1. Matrix of deterioration levels according to their types

Deteriorations classes	Cracks (longitudinal and transversal)	Potholes	Pull-out
A	0	0	0-1
B	1-2	1	2-3
C	3-10	2-4	4-5
D	11-20	5-10	6-10

The pavement surface condition indicator (SUI) for each km of paved road is determined according to the three classes of deterioration status, according to the following matrix:

Table 2. Surface deterioration matrix

Potholes	Pullout	Deteriorations			
		A	B	C	D
A	A	A	B	C	D
	B	A	B	C	D
	C	B	C	C	D
	D	B	C	D	D
B	A	B	B	C	D
	B	B	C	C	D
	C	C	C	D	D
	D	C	D	D	D
C	A	B	C	C	D
	B	C	C	C	D
	C	C	D	D	D
	D	D	D	D	D
D	A	C	D	D	D
	B	D	D	D	D
	C	D	D	D	D
	D	D	D	D	D

3.2 GIS method

Spatial technologies, such as geographic information systems (GIS), are particularly well suited to integrating roadway data and improving the use and presentation of that data for roadway management and operations by using spatial relationships to link geographic and geometric objects and events. To two different degrees, road management problems, such as pavement management, involve relationships between objects and events located in different spatial positions. Road networks span a large area and interact with various land features, including rivers, mountains, buildings, and other roads. Since the data used in the decision-making process have spatial components [5], space technologies appear to be an exciting alternative. Spatial technologies can improve the analysis of several transportation-related issues and improve the quality of the decision-making process.

In this regard, a complete database has been developed from the analysis of the environmental inspection results in three states:

- Creating a database that represents the following levels of road deterioration: potholes, pullouts, and cracks.
- Representation in the form of a matrix of deterioration subdivided into four levels: A, B, C, and D, to quantify and classify the deteriorations, to provide an image on the evolution of the structural and surface state, to identify the zones of equi-quality classified according to the levels of deteriorations [4], the linear interpolation of the data and translating them into SUI surface indicators.
- Modeling the SUI [2] them in a georeferenced thematic map.

In order to facilitate the reading of the data by the Arcgis software, all the levels of deterioration were transformed into quantitative variables. (Table 3)

Table 3. variable representation

Surface indicators classes	Reading in ArcGIS
A	1
B	2
C	3
D	4

The modeling of the surface indicators followed several steps on the ArcGIS software:

1. Georeferencing of the map and projection in the Moroccan coordinate system (Fig. 5)
2. Reading of the reduced database (Fig. 6)
3. Projection of the points on the map (Fig. 7)
4. Making the buffer (Fig. 8)
5. Classification by level of surface indicators (Fig. 9)
6. Realization of the thematic map (Fig. 10)

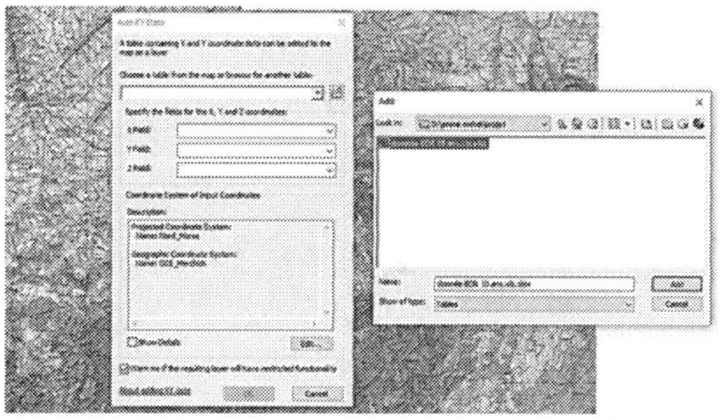

Fig. 5. Georeferencing of the map

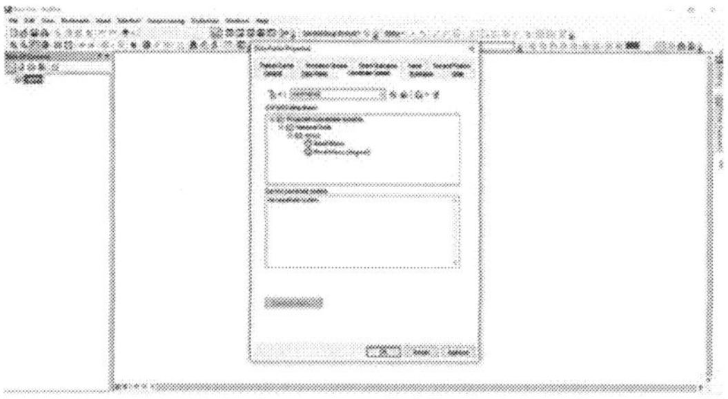

Fig. 6. Data reading

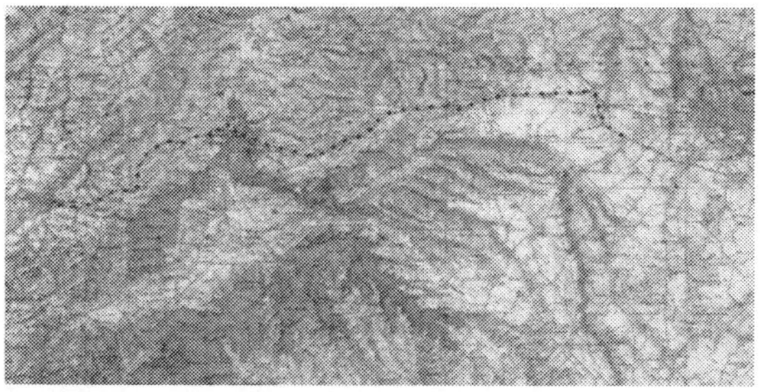

Fig. 7. Points projection

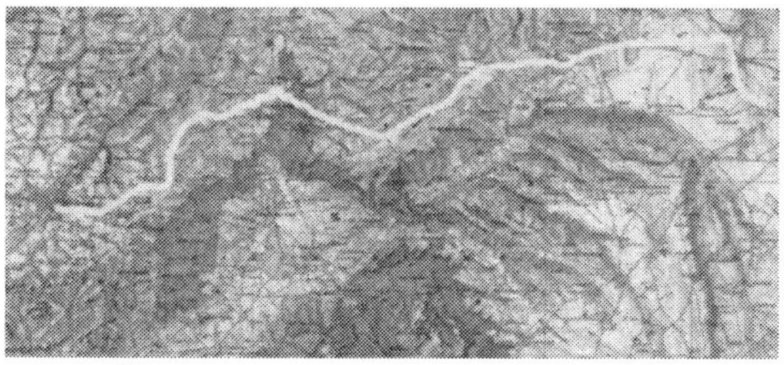

Fig. 8. Buffer

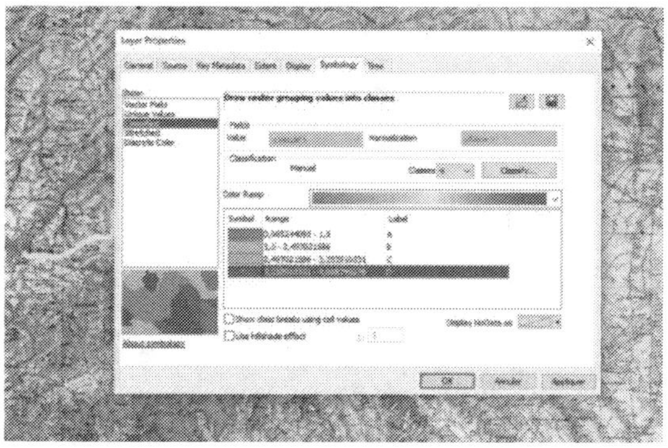

Fig. 9. Classification of indicators

4 Results and discussion

Regarding the situation of the national road number 06, inspections have been carried out in the presence of the administration on a monthly and annual basis. This road has undergone several repair interventions to protect users and ensure a level of comfort and quality on this road axis that has always experienced high demands due to heavy traffic and its attractive geographical location. These interventions include all measures taken by the contractor to ensure the desired level of service by reprofiling, localized resurfacing of shoulders, vegetation control and maintenance of drainage structures, as well as repair of pavement deterioration in order to maintain the level of service, including localized surface dressings and spot repairs including crack sealing and trimming for pavement body repairs.

These parameters are intended to ensure the viability of this roadway link which is open to two-way traffic on a 24-hour basis. According to the analysis of the results obtained in figure 9, the surface indicators reflect a low

https://doi.org/10.1051/e3sconf/202129804001

percentage in A+B, which means a substantial increase in the number of deteriorations. Despite the interventions which were recommended for this link, the map (Fig:10) represents a strong discontinuity between the various kilometers points as show an imbalance of surface continuation state on the 50 kilometers, the last kilometers while going towards Meknès are the most degraded (indicators of the state C and D).

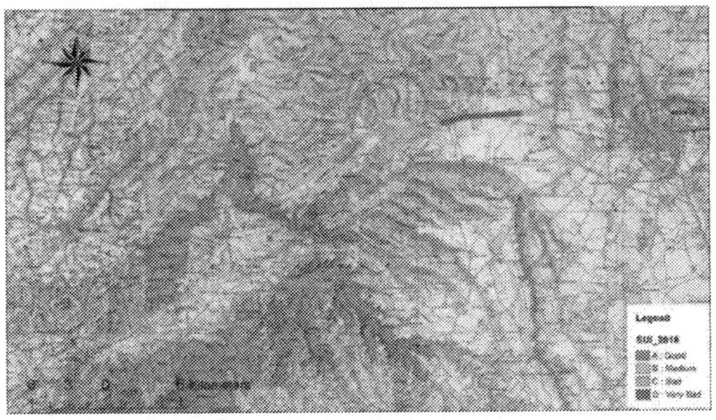

Fig. 10. surface indicators maps (SUI 2018)

5 Conclusion and perspectives

In this study, the objective was to model the deteriorations of flexible pavement in the short term (1 year) by applying the Moroccan method to survey the deteriorations, the classification, and the general reading of the general surface condition.

In this perspective, it is still necessary to replan the interventions by introducing a probabilistic prediction of the future deteriorations, by analyzing the evolution of the database, which is the object of the perspectives of the present research, by proposing a model of deterioration by adding other factors such as the evenness, the deflection and the variation of the structural state (STI: structural indicators) and their evolutions in time, in order to propose scenarios of maintenance and maintenance.

Acknowledgment

I want to thank my dear professor Mr. Toufik Cherradi for his relevant orientations, the National Centre for Road Studies and Research on behalf of Mr. Said El Karkouri, and the Moroccan Road Directorate on behalf of Mr. Ahmed Qachar for their collaboration and availability.

References

1. M.A.Mehdi, T.Cherradi, M.Quachar, A.Chigr, : Analysis of structural and surface deteriorations of a flexible pavement, based on inspection results using the MCA method. Materials Today Proceedings, Volume 45, Part 8, 2021, Pages 7538-7546.
2. H. A. Elhadi, "GIS, a Tool for Pavement Management," Master's of Science Thesis in Geoinformatics, Royal Institute of Technology (KTH), Stockholm, 2009. [Citation Time(s):1]
3. Lucky Sephumelele, Nkomoa, Sumaiya Amod Desaia, Khoboso Elizabeth Seutloalib, Kabir Yunus Peerbhaya, Timothy Dubec, Assessing the surface material quality of unpaved rural roads to understand susceptibility to surface deterioration. A case study of four rural areas in KwaZulu-Natal, South Africa Volume 112, August (2019), Pages 3-11.
4. Lucky Sephumelele, Nkomoa, Sumaiya Amod Desaia, Khoboso Elizabeth Seutloalib, Kabir Yunus Peerbhaya, Timothy Dubec, Assessing the surface material quality of unpaved rural roads to understand susceptibility to surface deterioration. A case study of four rural areas in KwaZulu-Natal, South Africa Volume 112, August (2019), Pages 3-11.
5. J. Neelam and P. K. Nanda, "Geographical Information System for Pavement Management Systems," Map Asia Conference, India, 2003.

Pedological characteristics of soils in the watersheds of Oueds Nekôr and Ghiss (Central Rif; Morocco)

Soukaina Ed Dakiri[1,*], *Hatim* Faiz[1], *Jaouad* Ouldchaib[1], *Said* El Moussaoui[1], *Hajar* El Talibi[1], *Hinde* Cherkaoui Dekkaki[1], *Taoufik* Mourabit[1], *Saida* Bouzid[2], *Brahim* Damnati[2], and *Issam* Etebaai[1]

[1]Laboratory of Research and Development in Engineering Sciences, Faculty of Science and Technology of Al Hoceima, Abdelmalek Essaadi University, Tetouan, Morocco.
[2]Environment, Oceanology and Natural Resources Laboratory, Faculty of Science and Technology of Tangier, Abdelmalek Essaadi University, Tetouan, Morocco.

Abstract. The objective of this work is the pedological characterization of soils in the watersheds of Oueds Nekôr and Ghiss. Several profiles were identified and sampled for Physico-chemical and sedimentological analyses (color index, magnetic susceptibility, pH, granulometry, water content, organic matter, and carbonates). The watersheds in question are roughly characterized by a soft and diversified lithology, a sparse vegetation cover, and a semi-arid Mediterranean climate. The soils in the Oued Nekôr watershed are of the silty type (> 50%), not very developed and relatively well-drained, and become well-drained stony in the steep and eroded areas. Soil pH varies between 7 and 8, reflecting low alkaline to alkaline soils. In the Oued Ghiss watershed, the sandstone-limestone or marl-schist soils are poorly differentiated in the steep slopes. The well-drained soils are gravelly, relatively alkaline, and show the beginning of rubefaction. Very locally on wooded flats, the soil profiles can become deep and well-differentiated due to colluvial accumulation. The dark brown eluvial A horizon is relatively rich in organic matter (3.6%). The well individualized illuvial B horizon has strongly rubbed concretions and is richer in organic matter (4.6%). The light beige C horizon corresponds to the strongly altered marl and limestone parent rock with low organic matter content (1.3%) and very low magnetic susceptibility values.

1 Introduction

The growing importance of environmental issues is leading to a renewed interest in soil heritage. The numerous hazards to which soils are subjected weaken their structure and affect their physicochemical and biological properties [1]. Water erosion is the primary cause of soil degradation in Morocco [2]. According to the High Commission for Water and Forests and the Fight against Desertification (HCEFLCD) (1996) [3], water erosion affects about 23 million ha of Moroccan land. In addition to the socio-economic damage and flood risks that threaten populations and infrastructure [4], this phenomenon reduces the hydraulic capacity of dams [5]. The north of Morocco, and particularly some regions of the Rif, are highly exposed to water erosion and siltation of water reservoirs because of the climatic aggressiveness, the rugged terrain, the tremendous lithological sensitivity (marl, marl-limestone, etc.), and the demographic pressure that knows the northern region [6], [7], [8], [9], [10], [11], [12], [13], [14].

The pedological characterization constitutes a preponderant step in all management strategies and agro-sylvo-pastoral restoration of such regions vulnerable to climatic contrasts and anthropic pressure [15]. The present work aims to characterize some pedological properties of some soil profiles collected in the watersheds of Oueds Nekôr and Ghiss.

2 Study sites

The study sites are located in the eastern part of the inner Rif in northeastern Morocco and are part of the Mediterranean coastal basins (Figure 1).

The Oued Nekôr watershed has an area of 773.43 km^2 and a perimeter of 170 km. It is limited to the north by the Mohamed Ben Abdelkrim Al Khattabi dam through which it flows into the full Al Hoceima and limited to the south by the watershed of Wadi Sebou and the west by the watershed of Wadi Ghiss, and that of Wadi Kert to the east. The Oued Ghiss watershed, located 25.5 km south of Al-Hoceima, has 586.66 km^2 and a perimeter of 132.52 km.

3 Materials and methods

Three soil profiles were collected in each watershed using the trenching method. Sampling was done every 5 cm. Wet color indices were determined using the Munsell chart. Sieving was done by wet sieving under running water to separate the coarse fraction from the fine fraction. The separation of the silty fraction from the clay fraction was done by sedimentometry. The pH of the soil was determined by a palliasse pH meter in a suspension of soil in distilled water. Bound water, organic matter, and

* Corresponding author: soukainaeddakiri@gmail.com

carbonate contents were determined by loss on ignition [16]. Magnetic susceptibility was measured directly with a portable susceptibility meter.

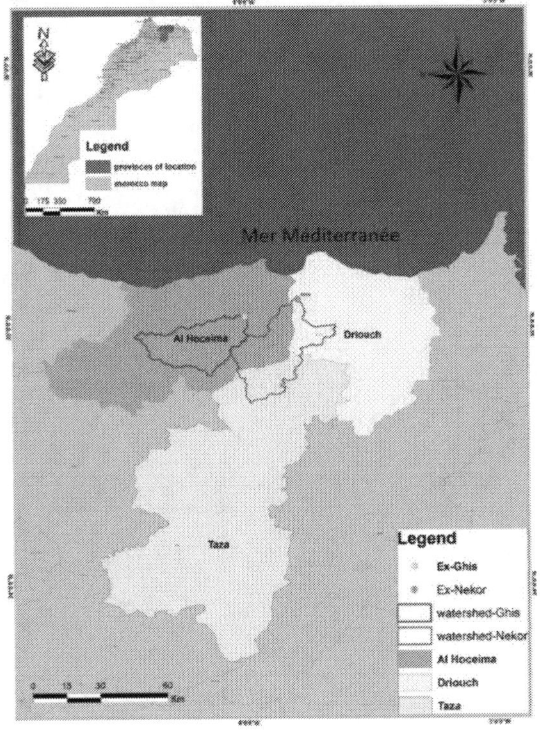

Fig. 1. The geographical location of Oueds Nekôr and Ghiss watersheds.

4 Results and discussion

4.1 Soil profiles of the Oued Nekôr watershed

Soil profile one was taken at the foot of a reforested almond tree slope on the left bank of Oued Nekôr. The other two profiles were taken from the fallow and sparse vegetation of the right bank.

The granulometric analysis reveals a predominance of the silty fraction for the different soil profiles studied in the Oued Nekôr watershed (Table 1). Profile 1 (80 cm long) is rich in coarse fractions (gravel and sand) at the level of the C horizon (35-75 cm deep). This fraction decreases significantly towards the surface in the B and A horizons, which become more affluent in silty fractions. The other two profiles (50 cm depth) are richer in silt with reasonably high percentages of clay (15% on average) in profile two between 10 and 30 cm depth. This distribution reflects a well-drained soil that has not evolved much overall. However, the beginning of structuring is underlined, as in profile 2, which shows a median accumulation horizon rich in clayey-silty fraction.

The pH values show a soil that is not very alkaline at the surface (pH=7.8) and becomes alkaline with depth in profile 1 (pH=8.41). In profiles 2 and 3, the alkaline soil is marked by a slight increase in pH with depth (Table 1). This evolution is mainly related to the low contribution of humic acids in connection with the low organic matter contents on the one hand and the precipitation of carbonates at the level of the soil horizons under semi-arid climate on the other hand.

Water content in the three profiles is very low and varies from 0.49% to 2.63%. Their variation with depth is marked by a decrease in profile one and an increase in profiles 2 and 3 (Table 1). This evolution of bound water content generally follows the granulometric texture of the soil horizons.

The organic matter content is also low and varies from 1.15% to 2.25%. Their evolution at depth is marked by their progressive decrease in profile one and their variability in profiles 2 and 3 (Table 1). This evolution is linked on the one hand to the predominance of coarse fractions that facilitate its rapid oxidation and, on the other hand, to the low contribution of organic matter sources of plant origin. The silty-clay fraction generally supports its distribution.

The carbonate content in profile 1 is higher than in profiles 2 and 3. Their variation at depth is marked in profile one by their increase as they approach the bedrock. These low carbonate values reflect a rather aluminosilicate nature of the bedrock and are generally related to the coarse fraction.

The magnetic susceptibility shows relatively low values for the three profiles studied (from 0.05.10-3 to 2.3.10-3 SI on average). Its evolution is marked by its decrease in profile one and its increase in profiles 2 and 3 (Table 1). Supported mainly by the fine silt-clay fractions, the magnetic susceptibility is conditioned by the distribution of ferric elements.

4.2 Soil profiles of the Oued Ghiss watershed

The soil profiles 1 and 3 were taken at the level of the sparse matorral flanks of the left bank of Oued Ghiss. Profile 2 was taken at the foot of a reforested almond tree and wheat cultivated flank of the right bank.

The granulometric analysis shows the predominance of the sandy-silty fraction in the three soil profiles studied, followed by gravel. Clay content is low in all three profiles and always remains below 8% (Table 2). The evolution with depth is marked by the increase of gravel on the one hand and the decrease of the sandy-silty fraction on the other hand in profile 1. While in profiles 2 and 3, there is an overall decrease in coarse fractions (gravels and sands) and an increase in fine fractions (silt and clay) from the middle horizon towards the bedrock (Table 2). This distribution reflects a well-drained and not significantly evolved soil. The measured pH of the soils in the Oued Ghiss watershed ranges from 7.35 to 8.17 and reflects a low alkaline to alkaline soil. The same conditions of oxidation of organic matter and precipitation of carbonates govern the distribution of soil pH in the Oued Ghiss watershed.

Table 1. Sedimentology and geochemistry of soil profiles of the Oued Nekôr watershed.

Profile	Depth (cm)	Munsell color	Gravel (%)	Sand (%)	Silt (%)	Clay (%)	pH	H_2O (%)	OM (%)	$CaCO_3$ (%)	χ (10^{-3} SI)
1	0-5	RYR 4/3	32,46	21,85	38,66	7,03	7,89	1,02	2,04	1,27	1,14
	5-10	RYR 3/4	49,21	16,99	31,64	2,16	7,78	1,51	2,26	2,02	0.36
	10-20	5YR 4/4	42,76	21,85	33,66	1,73	8,08	1,36	2,08	2,05	1,09
	20-35	2,5YR 4/4	29,71	26,84	40,14	3,31	8,12	1,44	1,85	3,10	1,22
	35-55	7,5YR 8/2	60,72	24,93	13,21	1,14	8,33	0,73	1,16	6,01	0,06
	55-75	10YR 6/2	74,13	18,75	6,53	0,59	8,41	0,49	1,03	5,72	0,08
2	0-10	10R4/4	21,66	26,80	45,79	5,75	8,02	1,77	2,10	1,48	1,54
	10-20	10R4/3	13,19	18,11	57,44	11,26	8,06	1,73	1,92	0,68	2,17
	20-30	10R3/3	19,45	15,51	46,48	18,56	8,07	2,43	1,96	0,01	2,33
	30-50	10R3/3	38,27	19,93	39,84	1,96	8,08	2,63	2,12	0,76	2,05
3	0-10	7,5YR5/3	46,89	32,91	19,24	0,97	8,22	0,64	1,45	0,37	0,70
	10-20	5YR4/3	22,86	29,11	44,53	3,50	8,32	1,78	1,72	0,45	1,66
	20-30	5YR5/4	16,79	28,30	51,00	3,91	8,39	1,48	2,20	1,07	1,01
	30-50	5YR3/3	14,18	27,10	54,47	4,25	8,64	2,18	1,99	0,89	1,25

BW: Bound Water; OM: Organic matter; $CaCO_3$: Carbonate; χ: magnetic susceptibility

Table 2. Sedimentology and geochemistry of soil profiles of the Oued Ghiss watershed.

Profile	Depth (cm)	Munsell color	Gravel (%)	Sand (%)	Silt (%)	Clay (%)	pH	BW (%)	OM (%)	$CaCO_3$ (%)	χ (10^{-3} SI)
1	0-10	2,5Y 4/3	7,01	53,58	34,19	5,21	8,03	0,52	1,47	0,59	0,41
	10-20	10Y 3/3	5,90	63,11	28,00	2,99	8,11	0,27	1.36	0,41	0,46
	20-40	7,5YR 4/3	26,36	45,04	26,10	2,51	7,35	0,74	1,46	0,56	0,27
	40-50	7,5YR 5/4	59,08	19,15	19,18	2,60	7,94	0,86	1,35	0,58	0,03
2	0 -10	5YR5/4	20,19	26,45	45,32	8,04	7,91	0,90	2,42	1,74	0,86
	10-20	5YR5/4	22,95	22,66	46,28	8,11	7,86	0,97	2,93	1,86	0,60
	20-40	5YR3/3	11,42	44,60	36,68	7,30	8,01	1,06	3,13	0,98	0,93
	40-60	5YR3/2	41,48	29,86	26,68	1,97	7,95	0,02	3,92	0,56	0,87
	60-80	5YR3/4	19,09	52,16	26,34	2,41	7,9	1,22	1,81	0,21	0,54
	80-100	7,5Y5/3	9,98	37,69	45,50	6,83	8,05	1,47	1,90	4,34	0,29
	100-120	10YR8/3	0,64	8,94	81,48	8,94	8,17	1,55	1,49	6,89	0,16
3	0-20	5YR4/3	32,39	28,99	33,22	5,41	8,00	0,42	2,54	0,42	0,47
	20-40	5YR4/4	33,85	40,68	22,34	3,13	7,77	0,56	1,44	8,64	0,52
	40-60	5YR5/2	32,88	35,82	28,11	3,19	7,86	0,52	0,45	0,26	0,31
	60-80	5YR5/4	16.90	38,97	39,49	4,65	7,88	0,47	1.84	2,45	0,45

BW: Bound Water; OM: Organic matter; $CaCO_3$: Carbonate; χ: magnetic susceptibility

The values obtained for the water contents at the level of the studied profiles present very low percentages and vary from 0.5% to 1.5%, reflecting a well-drained soil. The organic matter contents are low in profiles 1 and 3 and higher in profile 2. Their evolution in depth is marked by their increase at the average horizon B of profile two between 20 and 60 cm. The predominance of coarse fractions facilitates good drainage of the soil, favoring organic matter oxidation.

Carbonate contents are globally very low in the profiles studied. They are marked by a significant increase from 0.21% to 6.89% in profile two between 60 and 120 cm depth (Table 2). This variation is mainly related to the precipitation of carbonates in the upper horizons and the geochemical nature of the parent rock.

The magnetic susceptibility is low in the profiles studied. It varies from 0.15. 10-3 to 0.9.10-3 SI on average. The magnetic susceptibility decreases with depth in profiles 1 and 2 (Table 2). Its distribution remains in general anti-correlative with that of the carbonates.

5 Conclusion

The present work has allowed us to diagnose the state of evolution of some soil profiles at the Oueds Nekôr and Ghiss watersheds level. The contrasting climate with a semi-arid atmosphere, the soft and diversified nature of the geological substratum, and the abrupt relief are factors that control the pedological evolution of the soils at the level of the studied watersheds. The sedimentological and geochemical analysis of the different soil profiles provides information on well-drained silty soil, not significantly evolved, relatively alkaline, and poor in organic matter and carbonates. The granulometric analysis shows that the profiles are stony at the surface in the well-drained and leached areas and conversely stony with depth in the more stable profiles. The low content of organic matter supported by the fine fractions is related to the low biological productivity of the soil, which is due to the degraded state of the vegetation cover. Overall, the low carbonate and magnetic susceptibility contents reflect a marl-schist to sandstone-marly nature of the soils sampled in the Oueds Nekôr and Ghiss watersheds.

References

1. D. Touhtouh, Y. Moujahid, M. El Faleh, R. EL Halimi, J. Mater. Environ. Sci, 6 (12) 3582-3593, (2015)
2. D. El Hafid, B. Akdim, European Scientific Journal, 14(5) 259-285, (2018)
3. HCEFLCD. Plan National d'Aménagement des Bassins Versants, Cadre stratégique. Maroc, (1996)
4. E. Roose, M. Sabir, A. Laouina. Marseille : IRD, (2010)
5. A. Lahlou, IAHS Publ, 194 (1990)
6. M. Moukhchane, S. Bouhlassa. A. Chalouan, Sécheresse 3(9), 227-232, (1998)
7. M. Moukhchane, Bull. Réseau Érosion, 21, 255-266, (2001).
8. A. Sadiki, S. Bouhlassa, J. Auajjar, A. Faleh, J. Macair, Bull. Inst. Sci., 26, 69–79, (2004)
9. S. Ibrahimi, Thèse de Doctorat National. Université Abdelmalek Essaadi. FST Tanger, (2005)
10. S. Ibrahimi, B. Damnati, O. Radakovitch, K. Hassouni, B. Simon, Revista de la Sociedad Geologica de Espana 18(3-4), 217-224, (2005)
11. B. Damnati, S. Ibrahimi, O. Benhardouze, O. Radakovitch. M. Moukhchane, Africa Geoscience Review, 17(3), 215-223, (2010)
12. S. Ed-Dakiri, Mémoire de Master, Université Abdelmalek Essaadi. FST Al-Hoceima, (2019)
13. H. Faiz. Mémoire de Master, Université Abdelmalek Essaadi. FST Al-Hoceima, (2019)
14. M. Moustakim. Thèse de Doctorat National Université Abdelmalek Essaadi. FST Tanger, (2021)
15. I. Etebaai. Habilitation Universitaire, Université Mohamed Premier. Faculté des Sciences. Oujda, (2021).
16. W.E.J. Dean, J Sed Petro, 44(1), 242-248, (1974)

Geochemical and mineralogical characterization of coastal sediments in Al-Hoceima Bay (Central Rif, Morocco)

Mustapha Lamgharbaj[1,*], *Rachid* El Ansari[1], *Hajar* El Talibi[1], *Said* El Moussaoui[1], *Hinde* Cherkaoui Dekkaki[1], and *Issam* Etebaai[1]

[1]Laboratory of Research and Development in Engineering Sciences, Faculty of Science and Technology of Al Hoceima, Abdelmalek Essaadi University, Tetouan, Morocco

Abstract. This work aims to characterize the morphosedimentaty dynamics of the shores of Al-Hoceima Bay through geochemical and mineralogical analysis of surface sediments. The results reveal a direct relationship between marine hydrodynamic forcing on the one hand and detrital inputs from land on the other. Calcimetric analysis makes it possible to highlight the transport and deposition processes at the beaches of Al-Hoceima bay. The difference in the reasonably high carbonate contents in the western beaches (Isli, Cala Bonita and Quemado) (> 60%) and low in the beaches of the central and eastern zone (Sfiha, Souani, Salina and Lharch) (≈ 15%) are mainly due to the influence of coastal currents, littoral drift, and fluvial action. The higher levels of heavy minerals in Lharch beach (18%) are expressed by volcanic outcrops of Cap Quelates and Jbel Tamsamen. The low values at the Souani and Tayth beaches (7% on average) are due to erosion by the two types of coastal drift, NW-SE and NNE-SSW, which carry the sediments of schist origin brought by the Oueds Ghiss and Nekôr. In contrast, the reasonably high contents in the Sfiha beach (12%) are attributed to the accumulation of allochthonous sediments by NW-SE littoral drift which carries sediments of volcano-metamorphic origin from Rass Quelates.

1 Introduction

Soft or sandy Shorelines are fragile and vulnerable spaces where human pressure. Morocco coastal Mediterranean area is the subject of numerous particle size and mineralogical studies of surface sediments, mainly [1], [2], [3], [4], on the bay of Tangier, [5], on Al-Hoceima Bay, and [6], [7], for the Bay of Said. The morphology of the beaches is linked to the source of the sediments that feed them and the energy involved in the transport and deposition of sediments on the coastal fringe. Mineralogical and geochemical analysis of beach sediment identifies the relationship between transport, sedimentation, and sediment condition. The coastal area, especially the beaches of the Ghiss-Nekôr plain and Al Hoceima, have inspired man's action. They have undergone a wide range of facilities and activities, including port facilities, several tourist sites, waterworks on rivers (Abd Elkhatabi Dam on Oued Nekôr) etc. As a result, coastal operations are disrupted by these developments [6]. A superficial collection of 33 sand samples was carried out in different parts of the beaches of Al Hoceima Bay on both sides of the oueds, along 17 profiles perpendicular to the shoreline whose objective is to characterize the morpho-sedimentary sediment from the shores of this bay through geochemical and mineralogical analysis of the superficial sediments.

2 Study site

The Bay of Al Hoceima is located in the North of Morocco (Central Rif) on the southern facade of the Mediterranean, especially to the East of the city of Al Hoceima (Fig. 1). The bay is open to the North in an arch delimited by two rock formations: Ras Terf (Cap Quelates) to the East and Rass El Abed (Cap Maure) to the west. This bay extends over 40 km of coastline with more than 15 km of beaches. It is between 35°W 14' and 35°W 12' longitude and between 3°N 55' and 3°N46' latitude. It is bounded to the North by the southern part of the Alboran Sea and south by the Ghiss-Nekôr plain (city of Imzouren) (Fig. 1). Al Hoceima Bay is considered one of the largest bays in the Mediterranean. It has quite remarkable ecological and tourist potential. The hydrodynamics in this bay depends on the general dynamics of the Mediterranean. The marine zone of Al Hoceima is characterized by the anticyclonic gyre, which is oriented East-West [8]. The most frequent swells in the study area originate from WNW [8]. Those in the NE to ENE area are primarily to be considered. The NE swells cause coastal transit to the west of this bay. The Ghiss-Nekôr plain is made up of fillings of the heterogeneous alluvial quaternary (sand, gravel, and conglomerates) ranging from the Pliocene to the present [9], [10]. The study area is characterized by an alternation of several different morphological units from one sector to another.

[*] Corresponding author: lamgharbaj123mustapha@gmail.com

Rocky cliffed coasts are found in Capes, sandy coasts in the central part, and two largemouths (Ghiss, Nekôr).

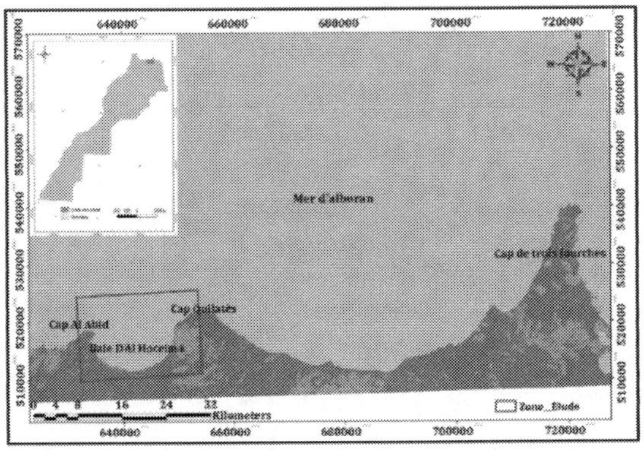

Fig. 1. Location of Al-Hoceima bay.

3 Materials and methods

3.1 Sand sampling

Thirty-three surface sand samples were collected using buckets during November and December 2019 and January 2020 and stored in transparent and airtight plastic bags. These sampling points are broken down into 17 profiles themselves into four sectors (Fig. 2). The processing and experimental sample operations were carried out at the laboratory of the Faculty of Science and Technology of Al Hoceima.

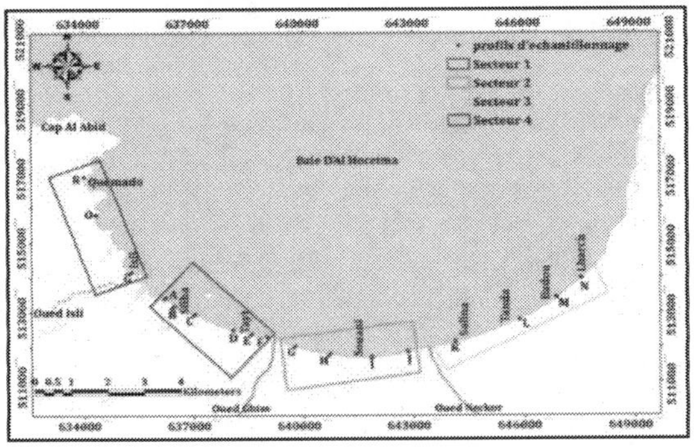

Fig. 2. Sector-based sediment sampling patterns at Al Hoceima Bay beaches.

3.2 Calcimetric analysis

We used the calcimetry method: the sand sediment in the mesh is less than 2 mm is ground and then dried. 0.1g of this crushed sediment, which is cold and hot-pressed with hydrochloric acid (HCl) above a stirrer, is taken. Excess acid is titrated by sodium hydrolysis (NaOH). This study measures the volume of carbon dioxide (CO_2) released by a sediment sample. This analysis makes it possible to know the amount of calcite present in the sediment. The calculation of the percentage of CaCO3 is based on a control test of pure calcite. The following formula obtains the percentage of carbonates:

$$\%CaCO3 = \frac{(Ve2 - Ve1)}{(Vc2 - Vc1)} \times 100 \qquad (1)$$

Vc2-vc1 is derived from the pure calcite (control) reading, and Ve2- Ve1e is the volume of CO2 from the sand sample released in the reaction.

3.3 Heavy minerals analysis

The mineralogical study method is carried out according to the experimental protocol proposed by Parffenoff et al. (1970). This technique consists of several steps and which are respectively:
- Preparation of samples,
- The separation of minerals,
- Assembly of minerals,

According to their density, the minerals constituting the sands are conventionally divided into two main groups: Heavy minerals with a density greater than 2,89 (Zircon, Amphibole, Olivine, and tourmaline, etc.) and light minerals with a density of less than 2.89 (quartz and feldspar). The preparation of the scales was carried out following a series of treatments (Wet sieving of the sediments under a net of ordinary water by a sieve of mesh 400 µm embossed in a sieve of 50 µm, the sands are decalcified by HCl cold then hot and Drying in the oven for 24 hours under a temperature of 105°C). Density separation was performed using 2.89-density bromoform (CHBr3) as a separator [11]. Heavy mineral blades were mounted between blades and slats using Canadian balm (two blades for each sample) on a hot plate to soften the viscous material after a series of treatments (sands are decalcified by HCl cold then hot).

4 Results and discussion

4.1 Contents of carbonates

Carbonate levels in the coasts of Al Hoceima Bay vary widely from the beach to beach and range from 9 to 70% (Fig. 3; 4). Carbonate levels in Sfiha range from 9.2% to 15.4% and are marked by a slight increase from west to East. The same trend is recorded at the Tayth beach, increasing from 13 to 14.50% eastward. From the right bank of the Nekôr mouth to the East of the Lharch beach, carbonate levels gradually increase by about 16 to 26%. The three beaches Quemado, Cala Bonita, and Isli, have the highest carbonate content (60% to 70%). The transverse evolution decreases the carbonates from the top to the bottom estran (Fig. 3; 4).

The level of $CaCO_3$ present in the sand samples tells us about the levels of calcite of geological and biogenic origin. Based on the longitudinal and transverse evolution analysis along the beaches of Al Hoceima bay, we found a direct relationship between marine hydrodynamic forcings on the one hand and detrital inputs on the other.

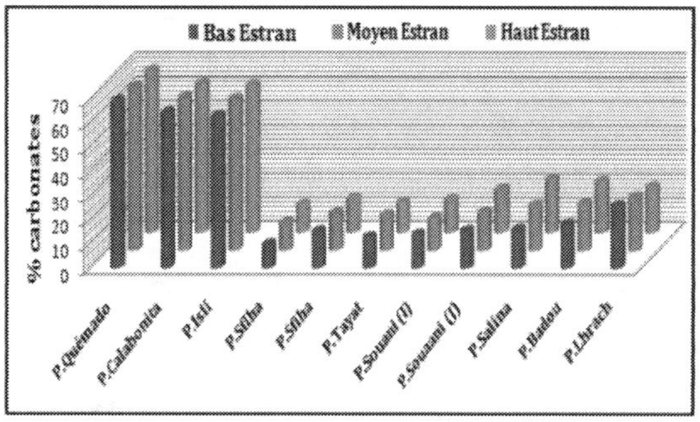

Fig. 3. Carbonate content of sand at the beaches of Al Hoceima Bay.

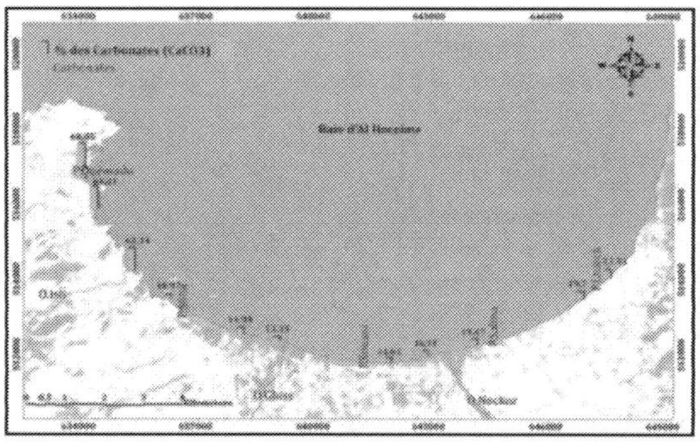

Fig. 4. Average Distribution of Sand Carbonate Levels at Al-Hoceima Bay Beaches.

In general, the results of calcimetry make it possible to highlight the transport and sedimentation processes at the bay beaches. Indeed, the difference in carbonate levels between the West beaches (Isli, Cala Bonita, and Quemado) and the central area of the bay (from Sfiha beach to Lharch) is mainly due to the influence of coastal currents, coastal drift, and fluvial action.

Carbonate inputs are of low biogenic origin (shells) and mainly detritic due to wave agitation on calcareous cliffs. This allows us to conclude that sand sediments in this area are generally Aboriginal due to local erosion of the cliffs and drainage of the hinterland that provides a high percentage of carbonates > to 65%. The low percentages thus are due to two main factors; the first is the marine erosion of the cliffs of Rass Tarf, which are volcanic origin formations devoid of calcite. The coastal drift of the NW-SE swell moves eroded particles to form the sediments of its beaches. The second is the river action of the Ghiss and Nekôr.

4.2 Heavy mineral contents

The sands' heavy mineral contents range from 6% to 18% at beaches in Al Hoceima Bay (Fig. 5). The high proportions were noted at the Lharch beach with 18%, followed by the Sfiha beach with 12% and then Quemado with 10%. The lowest levels were found at the mouth of Oued Ghiss (east of Tayth beach) and Souani beach with 8% and 6%, respectively (Fig. 5).

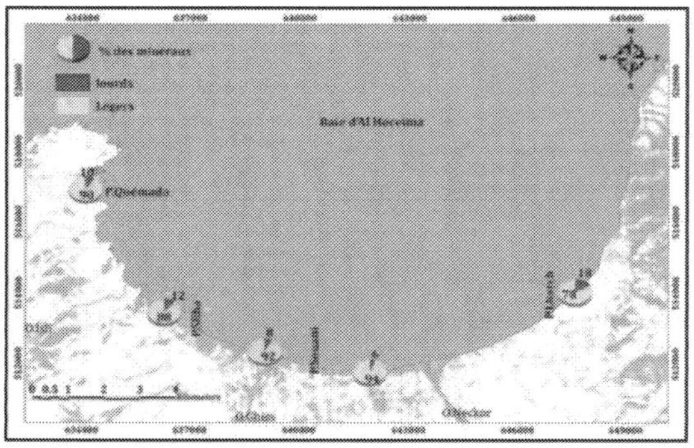

Fig. 5. Percentage of heavy and light minerals in sandy sediments at the beaches of Al Hoceima Bay.

The mineralogical analysis was carried out on five samples of the sandy fraction at different bay beaches. The higher quantity in Lharch beach is expressed by volcanic outcrops in Cape Quelates and Jbel Tamsamen on the one hand and may come from eroded sediments at the Nekôr mouth on the other. At the Souani and Tayth beaches, the low values in heavy minerals are due to erosion by the two types of coastal drift, NW-SE and NNE-SSW, which load sediments of shale origin brought by the Ghiss and Nekôr Oueds. Relatively high heavy mineral contents in Sfiha Beach (12%) with the presence of olivine in some beaches (Lharch and Sfiha) are attributed to the accumulation of allochthone sediments of blackish hue by NW-SE, which supports the sediments of metamorphic volcano origin of Rass Quelates.

5 Conclusion

As a result of this work, we were able to characterize the dynamics of surface sediments along the shores of Al-Hoceima bay. The amount of heavy minerals varies along the shoreline of Al-Hoceima Bay. It is marked by the high value (18%) in the Lharch range. This is due to the presence of volcanic outcrops in Cape Quelates. Calcimetry showed that the carbonate levels in the sediments studied are variable at this shoreline; the lower values are found in the central part of this bay (6%). This is due to the dominance of shale geological formations in the Ghiss and Nekôr catchment innervated in this coastline's diet.

References

1. A. Boughaba, Thèse de Doctorat, Fac. Lettres, Nantes, (1992)
2. A. El Arrim, Thèse de Doctorat, Université. Abdelmalek Essaadi, Fac. Sci. Tech. Tanger, (2001)
3. I. El Hatimi, M.Achab, B. El Moumni, Bull. Inst. Sci., 24, 49-58, (2002)
4. M.Achab, B. El Moumni, A. El Arrim, J.M. Gutierrez Mas, Bull. Inst. Sci, (27), 00-00, (2005)
5. B. Gensous, M. Tesson, Mines, Géol. et Energie., 46, 127-134, (1979)

6. M. Bouabdallah, Thèse de doctorat. Paris Est, (2008)
7. N. Bellahbib, H. Rezqi, M. Oujidi, S.Bengamra, LARHYSS Journal, (24), 19-40, (2015)
8. A. Khouakhi, Thèse de doctorat, université Mohamed, (2013)
9. A. Salhi, Thèse de doctorat, Université Abdelmalek Essaadi, Faculté Sciences, (2008)
10. N. Iouzzi, Mémoire DESS, Université Med V, (2008)
11. A. Parfenoff, C. Pomerol, J. Tourenq, Sédim Pétrol, Vol 34(4), 830-847, (1970)

 https://doi.org/10.1051/e3sconf/202129805001

Heavy metal contamination in surface water of Mohammedia wetland, Morocco

Halima Jounaid[1,*], *El Mehdi* El Hachimi[1], *Nihad* Chakri[1], *Toufik* Remmal[1], *Btissam* Elamrani[1], and *Fouad* Amraoui[1]

[1]Hassan II University of Casablanca, Faculty of Sciences Ain Chock, Department of Geology. Km 8 Route d'El Jadida, Casablanca 20150, Morocco.

Abstract. Assessing heavy metal concentrations in wetlands and identifying sources of metal contamination are critical steps in protecting wetlands. Using seven sampling stations, we evaluated the spatial variation of some heavy metal concentrations in surface waters of the Mohammedia wetland and a selected segment of the El Maleh wadi that feeds the wetland. Field observations made it possible to identify and map the pollution discharged into El Maleh wadi, carrying heavy metals into the wetland. The analyses showed that the concentrations of lead, manganese, and cadmium far exceed the thresholds relating to fish life in 100% of sampling stations. All the evidence points to heavy metals contamination of surface water in Mohammedia wetland, generated by industrial activities, wastewater discharges, and leachate from the old landfill located upstream.

1 Introduction

Water quality is a potential factor in supporting biodiversity [1]. Water quality degradation can seriously affect wetland ecological functions [2-4]. Heavy metals from anthropogenic sources are considered one of the most common types of pollutants [5]. Heavy metals are widely present in rivers and serve as important indicators of environmental water quality [6]. Riverine input and runoff from land carry pollutants into estuaries, thus impacting biotic communities and sensitive habitat areas [7].

Mohammedia wetland, located in the eponymous city in the west of Morocco, is a Ramsar site corresponding to El Maleh Wadi's estuary and its flood plain [8]. The wetland mainly consists of some marshes fed by rainfall and El Maleh Wadi floods, a permanent water body supplied by seawater during high tide, and another permanent body of water fed by groundwater [9]. In the latest decades, the urban development of Mohammedia city has led to the release of wastewater into El Maleh Wadi. Several industries (mainly an oil refinery and coal-fired power plant) surrounding the wetland also pose a severe threat. Some studies are available on the surface water quality of Mohammedia wetland [10-11] but have not assessed heavy metals in surface water.

In this paper, we highlight the spatial variation of some heavy metals in mohammedia wetland surface water. The goal is to investigate heavy metal concentrations in different areas and identify heavy metals contamination sources of Mohammedia wetland as the first step for its protection and Restauration. Physico-chemical analyses were carried out on five heavy metals. The concentrations obtained for each parameter are compared to the thresholds tolerable for fish life [12]. The objective is to verify the ability of Mohammedia wetland to offer natural conditions for biodiversity.

2 Materials and Methods

The sampling network was designed to cover sites that reasonably represented El Maleh Wadi's water quality, taking into account discharges identified during the field observations and impacting Mohammedia wetland water quality. Sampling station 11 is chosen as the reference station given its location far from anthropogenic disturbances. The surface water samples were made in June 2019. All samples were collected in double, in new polypropylene bottles rinsed with deionized water, then stored at 4°C in the dark and transported on the same day to the laboratory. The analyses were performed according to Rodier [13] in the Laboratory "Geosciences Applied to the Engineering of the Development" (GAIA) in Hassan II University of Casablanca – Ain Chock Faculty of Sciences. Lead (Pb), Manganese (Mn), cadmium (Cd), Chromium (Cr), Zinc (Zn), and copper (Cu) were measured at seven selected points (1-3, 5, 6, 9-11) (Fig. 1).

* Corresponding author: jounaidhalima@gmail.com

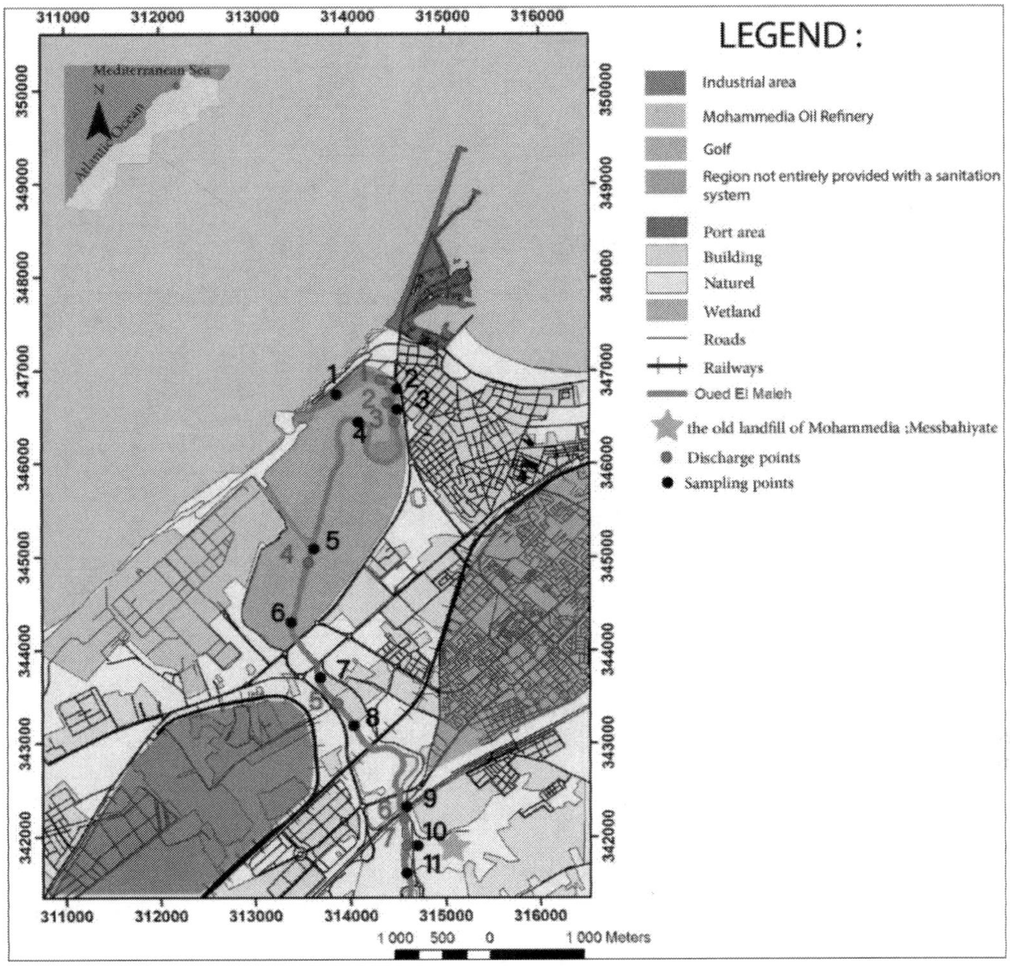

Fig.1. Study area and water quality sampling sites.

3 Results and Discussion

The mean concentrations of heavy metals are summarized in table 1. The Analyses carried out show that, apart from Zn, all other heavy metals were detected in levels exceeding maximum values tolerable for fish life. The concentrations of lead in the study area are significant. Due to urban processes, Pb and other metals are regularly discharged into fields, water, and soils through sewage sludge, urban runoff, automobile exhaust, and gasoline containing Pb [14-15].

The highest concentration of lead recorded in the sampling station (S10) (Fig. 2a) may originate from the old landfill of Mohammedia city. During the field observations, we noted in this area leachate seepage into El Maleh wadi. Analyses carried out on the leachate from Mohammedia landfill have shown lead concentrations up to 690 µg/L [16]. Sample station ten also shows the highest concentration of manganese (Fig. 2b). Wastewater and industrial pollution are also responsible for surface water contamination by manganese [17]. Indeed, S9, S5, located near wastewater discharges in El Maleh wadi, show significant these parameters concentrations. Concerning cadmium, all levels detected far exceed the maximum allowable for fish life (Fig. 2c). Except in station 10, downstream concentrations are more important than those recorded upstream. Station S1, located in the estuary, shows the highest value (100.5 µg/L). The affinity of Cd can explain this with chlorine in a salty environment. This affinity results in the desorption of Cd from the particulates towards the dissolved phase [18-20]. The significant contamination of sites S1 and S3 by Cd could be related to phosphate fertilizers runoff [21] which could be used in the Mohammedia golf that covers 27 ha in this area. According to N'Gessman [21], some phosphate fertilizers have cadmium concentrations that can be about 300 times higher than the natural content of soils. Copper was detected in only three samples. The station (S10), located downstream from the landfill, records a value far exceeding the threshold of this parameter (Fig 2d), attesting to possible contamination of surface water by the leachate.

Table 1. Parameters measured in the study area during June 2019

Sampling station	S1	S3	S5	S6	S9	S10	S11	Thresholds Fish life [12]
Pb µg/l	32.5	161.5	64.5	0.1	64.5	226	53.94	20
Mn mg/l	0.89	0.92	1.02	1.01	0.96	1.40	0.10	0.1
Cd µg/l	100.5	78	60	48.5	53	77.5	47.5	5
Zn mg/l	0.27	0.23	0.022	0.07	0.04	0.32	0.15	1.3
Cu µg/l	-	13.5	-	-	-	137	13	40

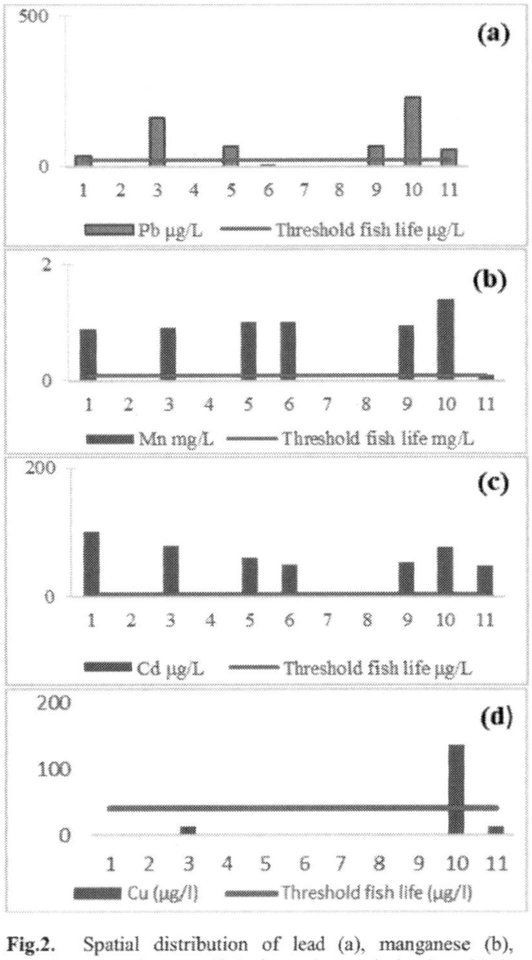

Fig.2. Spatial distribution of lead (a), manganese (b), cadmium (c) and copper (d) in the study area during June 2019.

4 Conclusion

In this study, industrial discharges, domestic sewage, and the old landfill of Mohammedia city were identified as potential pollution sources in the study area. Mn, Cd, and Pb are the major heavy metal contaminants in El Maleh Wadi and Mohammedia wetland surface water. The spatial variation of heavy metals concentrations shows that pollution discharged in El Maleh Wadi reaches Mohammedia wetland and may affect aquatic communities [22]. To prevent the ecosystem health risks, there is a critical need for implementing effective measures to reduce pollution in El Maleh Wadi, such as treatment of the old landfill leachate and treatment of domestic and industrial wastewater before discharging it into El Maleh wadi.

References

1. W. Duan, B. He, D. Nover, G. Yang, W. Chen, H. Meng, S. Zou, C. Liu. Water Quality Assessment and Pollution Source Identification of the Eastern Poyang Lake Basin Using Multivariate Statistical Methods, Sustainability **8**, 133 (2016).
2. B.E.M. Keller, K. Lajtha, S. Cr.istofor. Trace metal concentrations in the sediments and plants of the Danube Delta, Romania, Wetlands **18,** 42–50 (1998).
3. Y. B. Wang, C.W. Liu, P. Y. Liao, J.J. Lee. Spatial pattern assessment of river water quality: implications of reducing the number of monitoring stations and chemical parameters, Environ. Monit. Assess. **186**, 1781–1792 (2014).
4. X. Jiang, S. Xu, Y. Liu, X.Wang. River ecosystem assessment and application in ecological restorations: A mathematical approach based on evaluatin gits structure and function, Ecol. Eng. **76**, 151–157 (2015).
5. K. Pan, W.X. Wang. Trace metal contamination in estuarine and coastal environments in China, Sci. Total Environ. **421**, 3–16 (2012).
6. K. Peters, M. Bundschuh, R. Schäfer. Review on the effects of toxicants on freshwater ecosystem functions, Environ. Pollut. **180**, 324–329 (2013).
7. M. J. Kennish. Environmental threats and environmental future of estuaries, Environ. Conserv. **29** (01), 78–107 (2002).
8. M. Dakki, M.A. El Agbani. A. Qninba. Zones humides du Maroc inscrites jusqu'en 2005 sur la liste de la Convention Ramsar, Bull.Ins. Scien. Rabat, série générale **7**, 226 (2011).
9. A. Serghini, A. El Abidi, S. El Blidi, M. Fekhaoui. Typologie spatio- temporelle et fonctionnement hydrologique du complexe des zones humides de Mohammedia (Maroc), Bulletin de l'Institut

Scientifique, Rabat, Section Sciences de la Vie **35**, 61-65. (2013)

10. A. Serghini. Diagnostic du complexe zones humides de Mohammedia : étude qualitative et quantitative des composantes physique (eau – sédiment) et biologique de l'écosystème. Univ. Mohammed V, Rabat, Ph D, Thesis, 178 p (2003).
11. A. Serghini, M. Fekhaoui, A. El Abidi, S. El Blidi, R. Ben Akkame. Caractérisation hydrochimique d'un site Ramsar : le complexe de zones humides de Mohammedia (Maroc), Bull. Inst. Scien. Rabat, section Sciences de la Vie **32**, 133-145 (2010).
12. Ministry of Energy, Mining, Water and Environment. Available online:
    http://www.environnement.gov.ma/fr/lois-et-reglementations/normes?id= 1010.
13. J. Rodier, *L'analyse de l'eau, Eaux naturelles, Eaux résiduelles, Eaux de mer* (Dunod, 2009).
14. K.C. Cheung, B.H.T. Poon, C.Y. Lan, M.H. Wong. Assessment of metal and nutrient concentrations in river water and sediment collected from the cities in the Pearl River Delta, South China, Chemosphere **52**,1431-1440 (2003).
15. A. A. Elnazer, S. A. Salman, E. M. Seleem, E. M. Abu El Ella. Assessment of Some Heavy Metals Pollution and Bioavailability in Roadside Soil of Alexandria Marsa Matruh Highway, Egypt, Int. J. Ecol, **2015** (2015).
16. S. Souabi, K. Touzar, H. Chtioui, F. Khalil, K. Digua, M. Tahiri. Problématiques du chrome et du plomb dans les lixiviats des décharges publiques des villes de Mohameddia et de Fès, Déchets sciences et techniques **58**, 37-44 (2010).
17. World Health Organization, *Manganese and its compounds : environnemental aspects* (Tech. rep. World Health Organization, 2004).
18. A.C. Bourg. Trace metal adsorption modelling and particle-water interactions in estuarine environments, Cont. Shelf Res.**7**, 1319–1332 (1987).
19. A. Turner, G. Millward, A. Bale, A. Morris, Application of the K D concept to the study of trace metal removal and desorption during estuarine mixing, Estuar. Coast. Shelf Sci. **36**, 1–13 (1993).
20. M. Waeles, R.D. Riso, P. Le Corre. Seasonal variations of cadmium speciation in the Penzé estuary, NW France. Estuar. Coast. Shelf Sci. **65**, 143–152 (2005).
21. Y.M. N'Guessan. Dynamique des éléments traces dans les eaux de surface des bassins versants agricoles de Gascogne, National Polytechnique Institute of Toulouse, Toulouse, Ph. D. Thesis, 253 p (2008).
22. R.B. Schäfer, M. Bundschuh, D.A. Rouch, E. Szöcs, P. C von der Ohe, V. Pettigrove, R. Schulz, D. Nugegoda, B.J. Kefford. Effects of pesticide toxicity, salinity and other environmental variables on selected ecosystem functions in streams and the relevance for ecosystem services, Sci. Total Environ. **415**, 69-78 (2012).

Modeling and Mapping of coastal aquifer vulnerability to seawater intrusion using SEAWAT code and GALDIT index technique: the case of the Rmel aquifer – Larache, Morocco

Mohamed Jalal El Hamidi[1], *Abdelkader* Larabi[1,*], and *Mohamed* Faouzi[1]

[1]Regional Water Centre of Maghreb, EMI, Mohammed V University in Rabat, Rabat, Morocco

Abstract. The study area of Rmel-O. Ogbane aquifer, located in the north of Morocco, currently faces major water challenges related to the sustainable management of water resources. Climate change and Sea-Level-Rise can increase the risks and costs of water resources management and impact water resources' quantity and quality. Hence, for planning and management, an integrated approach is developed for linking climate models and groundwater models to investigate future impacts of climate change on groundwater resources. Climate projections show an increase in temperature of about 0.45 °C and a reduction in precipitation of 16.7% for 2016-2050. Simulations of seawater intrusion corresponding to various combinations of groundwater extraction predicted climate change and sea-level-rise show that the area will be contaminated on the NW sector of the coastal part. The toe would reach about 5.2 km inland and intrude on high salinity (15–25g/l). Beyond these zones, the contamination of the aquifer will be limited. Moreover, these results were confirmed by the application of the GALDIT method. They reveal that the fringe littoral areas of the aquifer are the most affected by seawater intrusion, with a high risk in the north-western part of the study area.

1 Introduction

Studying the effect of Climate Change (CC) on groundwater resources is essential to water planners and managers because it may change underlying water management conditions [1, 2]. The impact on the groundwater system may be described by referring to its quantity, quality, and functionality. Groundwater recharge is controlled by soil infiltration, land cover, and precipitation amount. Under CC, the variability in the precipitation regime (yearly precipitation amount concentrated in a few precipitation events), combined with land-cover degradation, significantly after groundwater recharge, will be causing a deficit of water balance in the aquifer as well as a degradation of the freshwater quality by seawater intrusion on the coastal plain.

The present study focuses on the Rmel-O. Ogbane coastal aquifer (ROOCA) is intensively exploited. This aquifer is located in the North-West part of Morocco and is very well known for its essential role in the region's industrial, economic, and social development. The paper is based on analyzing the impacts of CC on water resources in the study area based on a 3D groundwater flow modeling and GALDIT method mapping.

The activities can be divided into six main steps: (1) First, analyzing some climate data from available weather stations and collecting future predicted CC datasets with temperature and precipitation variables regarding Rmel-O. Ogbane area in Morocco from 2017 to 2050 based on climate modeling. Indeed, a better understanding of trends can be recognized from observed and forecasted precipitation time series and temperature data in the region, which should be related to CC. More exactly, what impacts these changes will have on the aquifer recharge; (2) based on this climate data (evapotranspiration and rainfall), they are used to estimate recharge for the period 2017-2050; (3) organizing all these data in standard formats stored in a GIS database; (4) creating GIS database better to understand the hydrogeological structure and geometry of the reservoir; (5) identifying and evaluating the spatial distribution of regionalized variables on reservoirs and groundwater resources using geostatistical analysis; (6) analyzing the role of human activities on CC and its impacts on groundwater resources in this area. More specifically, heavy abstraction rates and impacts these changes will have on groundwater level and water quality, and finally (7), both a two-dimensional GALDIT method and a three-dimensional groundwater model are used to identify coastal aquifer vulnerability to seawater intrusion. The developed model considers the increase of groundwater abstractions to satisfy the water demand of urban centers of both Larache and Ksar El Kebir cities. It is capable to better understand the hydrodynamic functioning of the aquifer system by assessing the components of the groundwater mass balance to determine the spatial distribution of the hydrogeological parameters. The model and the method developed will be used as a management tool to assist in

* Corresponding author: larabi@emi.ac.ma

 https://doi.org/10.1051/e3sconf/202129805002

planning and water resources management of the aquifer system.

In the upper part of the chart (Fig. 1), tasks assemble several climate data sets for current and future predicted conditions derived from climate modeling, which are used to simulate recharge using the Thornthwaite method. This monthly recharge is then used in a three-dimensional groundwater model (lower part of the chart) to simulate transient saturated variable-density groundwater flow to assess and predict the CC Impacts on groundwater resources and socio-economic vulnerability.

Moreover, the application of the GALDIT method, which consists of mapping the aquifer vulnerability to seawater intrusion (SWI), has been conducted to locate the most affected areas by seawater intrusion.

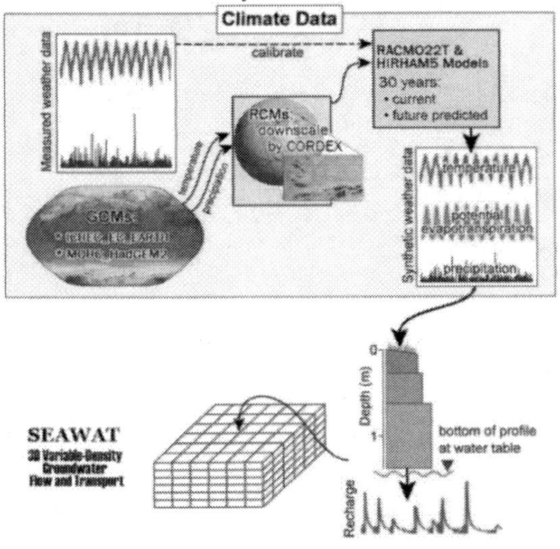

Fig. 1. Flow chart of tasks showing coupled Regional Climate Models to groundwater Models.

2 Background of the study area

2.1 Location of the study area

The Rmel-O. Ogbane aquifer is located in the Low-Loukkos basin immediately at the south of Larache city (Fig. 2). It extends over an area around 305 km^2 and bordures the Atlantic Ocean to the West by the salty level, over a 20 Km band. It is limited by slope gray silts on the left bank of Loukkos wadi and to the East by Mio-Pliocene marl outcrops at the South-East, and the south by the ascended bedrock constitutes a water divide line between the Rmel-O. Ogbane aquifer and the Dradere-Soueire area [3].

The annual average rainfall in the Lower-Loukkos basin reaches more than 1120 mm in the 60th and declines to 378 mm/y in 2016. The seasonal variation using 12-month moving average, the monthly values indicate a decrease over the last five decades after the most important intensity during the 1960s. The variation shows a clear seasonal variability. The monthly values of the temperature in the Lower-Loukkos basin indicate an increase during the last three decades since the 1976's. The impact of CC in the Lower-Loukkos basin cause recurrent droughts, and decreases in recharge directly affect the groundwater level in the aquifer, where it's continuing to decrease.

This is associated with excessive over-pumping rates related to the sustainable management of water resources and the delivery of water services for domestic, agricultural, and industrial use. The aquifer supplies good groundwater quality that is easily accessible [4]. This favorable situation has increased pumping and caused environmental problems, such as water table decline and saltwater intrusion on the coast and the coastal plain of the study area. Inappropriate management of this coastal aquifer (overexploitation) may lead to saltwater intrusion into freshwater wells, destroying them as sources of fresh water supply.

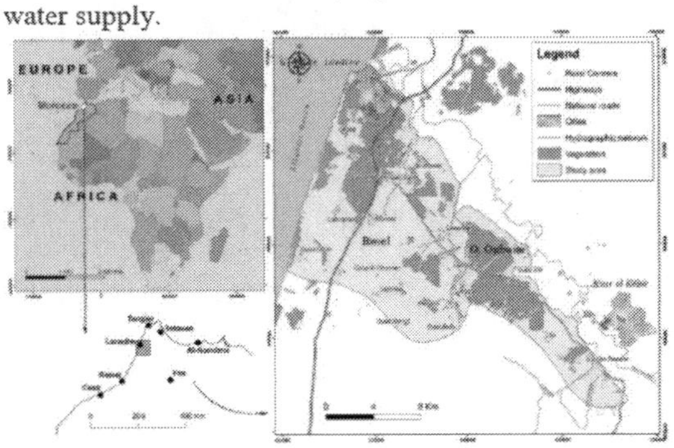

Fig. 2. Location map of the study area

2.2 GIS Database

A database was developed based on collected data from various studies ([5, 6, 7, 8]), offices, and organizations such as Direction of the Research and the Planning of Water (DRPE), Hydraulic Basin Agency of Loukkos (ABHL), Regional Office of Agricultural Development of Loukkos (ORMVAL), National Office of Drinking Water of Tangier and Kenitra towns (ONEE), Mohammadia School of Engineers (EMI) and 'Conseil Ingénierie et développement' (CID) which describes the surface water and groundwater resources (hydrology, geology, geometry, groundwater level, and groundwater quality, etc.).

These collected data from the beginning of the 1950s to the year 2016 were collected from the six main organizations mentioned above. All these collected data have been homogenized and stored in a database network of different types: relational geodatabase, geodatabase, and information layers' MONAROO Toolbar' [9].

2.3 Hydrogeology and Geostatistics

From the hydrogeological point of view, the Rmel-O. Ogbane aquifer consists of a very heterogeneous complex of fluviolacustrine formations. It consists mainly of Moghrebian shelly sandstones [10] surmounted by Quaternary sands and red silts more or less marly in the Rmel area and laterally evolves in pebbles and sandy silt in the Oulad-Ogbane area. In some places, an intermediate formation of clayey sands and sandy clays is intercalated between the two units of the Moghrebian and Quaternary and can constitute a semi-permeable screen isolating the two hydrogeological units.

A geostatistical model was realized, and some prediction maps were integrated into the GIS. The normality test and the analysis of trends were studied for each variable to select the appropriate Semivariogram and check the results using cross-validation [4].

3 Results: Numerical modeling of SWI

3.1 Conceptual model and boundary conditions

The model grid in the plan view is made up of 128 columns and 128 rows of 250 m (NS-direction) and 250 m (WE-direction) uniform spacing. Moreover, the model grid, vertically oriented, comprises three-layer construction that corresponds to the Rmel hydro-stratigraphy layers.

Since the model was designed to simulate SWI into both GW flow and solute transfer processes were active in coastal aquifers simultaneously, with boundary conditions taken into consideration. The boundary conditions were set as a constant head boundary around the Atlantic coast for the flow computation, where the cells have a persistent head (mean sea level). The initial head values are set to grid top elevation and turned on in the MODFLOW/SEAWAT program. Moreover, the SEAWAT program was used to simulate the model in a steady-state over 100 years to calculate the initial concentrations of seawater intrusion in 1961. This result will be assigned as the initial concentrations for the transient state. The cells along the Atlantic coast are given a steady concentration of 35 kg/m^3. Another no line was established through marl outcrops to the southeast of ROOCA and by the risen bottom to the south, which serves as a dividing line between ROOCA and the Dradere-Souiere aquifer. A flow boundary was established in the rest of the aquifer based on the observed data. The aquifer has, for the subsequent layers, internal hydrological stresses are present. The aquifer bottom has a no-flux limit, while the aquifer top has a rebound flood boundary (rain and irrigation return flow).

3.2 Calibration and validation

Figure 3 shows a good match between measurements and calculations of heads in the 24 observation wells within the modeled region. It also shows the effects of this calibration that the model calibration yielded the two sets of values are highly correlated, as seen by the map, which is very similar to the perfect correspondence, with an average correlation coefficient of 0.999, a mean error of 0.19 m, and a root mean square error of 0.849 m. The mean percentage difference is about 0.827 m.

The measured and estimated heads are relatively similar. The overall model is close, with ridges or headers that are almost identical. The lateral hydraulic conductivity of Layers 1, 2, and 3 as a result after model calibration varied from 4.10^{-4} to 8.10^{-7} m/s.

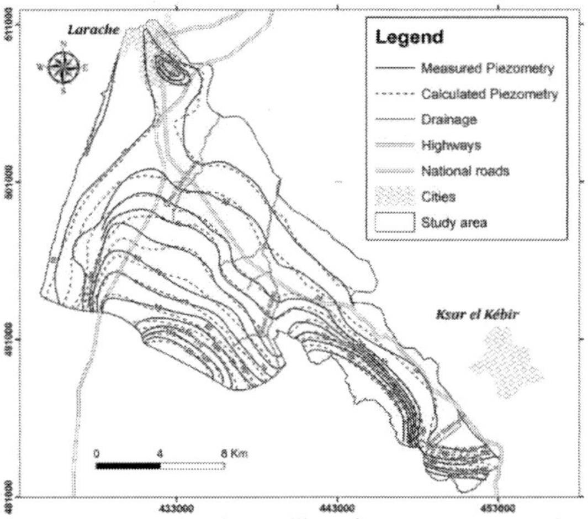

Fig. 3. Map displaying calibrated outcomes comparing calculated and measured steady-state heads (1961/62).

However, in a transient state, the model was calibrated from 1961/62 to the end of 2016. In complete 22 observed hydraulic head calculations were employed during calibration. The calculation is done for the goal date of 1963-2016 (53 years). When the fit between the observed and measured heads was reached, model calibration was terminated.

The simulation, in this case, tests piezometric surface variations over time. Hence, the parameters involved in the temporal equation can be calculated (either a particular yield or specific storage).

Aquifer boundary inflows (flux at the boundary, meteoric recharge) and outflows (withdrawals) were also studied. The goal was to replicate the piezometric oscillations found at the control points as accurately as possible during the measurement period. The system reference time is when GW level measurements were made, with the computed piezometric surface in steady-state conditions acting as the initial condition. The model is set up with the initial concentrations of SWI in 1961/1962. The simulation lasted a total of 20075 days, beginning in 1962 and ending in 2016.

This simulation makes use of wells that pumping water is mostly used for irrigation and DWS; irrigation started in 1981 and will last until 2016; withdrawals differ over time and are allocated based on piezometric variations. After calibration, the adopted basic storage was estimated to be between 0.1 and 4 percent. This distinction demonstrates adequate consensus between measured and estimated heads in various observation wells for the simulation time under consideration.

3.3 Forecasted numerical modeling

The simulation period is based on the groundwater management scenario established by the National Office of Drinking Water (ONEE) in 2016 and ranges from 2017 to 2040. This scenario is based on satisfying future drinking/industrial water demand of both cities Larache

https://doi.org/10.1051/e3sconf/202129805002

and Ksar El Kebir and keeping groundwater abstractions for irrigation constant under CC forcing scenario RCP4.5 and associated with SLR. The numerical model considers three vertical layers to represent the density variation accurately within the aquifer depth. The model was extended one to two kilometers into the Atlantic Ocean, where a constant head and constant salinity boundary was imposed, taking into account the predicted SLR at the case study. Figure 4 shows the results of the predicted drawdown and saltwater extent intruded into the freshwater of the aquifer by 2040.

Under this climate moderate scenario (RCP4.5) and SLR, as integrated into the numerical model, the results show that by 2040, there will be more decrease in hydraulic heads due to intensive pumping discharge to satisfy the water demand of the two urban centers. Indeed, the hydraulic heads will reach negative values around -20 m at the main ONEE well field area, where the drawdown will increase by 25 m, which is highly significant. The aquifer is also contaminated by SWI in the NW sector of the coastal part, in which the toe would reach about 5.2 km inland with an invaded area of about 31 km^2 and will be intruded with high salinity (15–25 g/l) in Layer 3. Consequently, seawater would reach seven observation wells (1534/3, 1535/3, 1536/3, 1380/3, 342/3, 1396/3, and 438/3) and four pumping wells (417/3, 419/3, 718/3, and 1737/3) gradually in 2040. The expected salinity, greater than 2g/l, will be already reached in 2032 and will continue to increase up to 6 g/l in 2040 (Fig. 5). Also, it indicates that the salinity is expected to rise by 2040 from 20 to 30 g/l at piezometer 1380/3, located at 1.5 km far away from the coast.

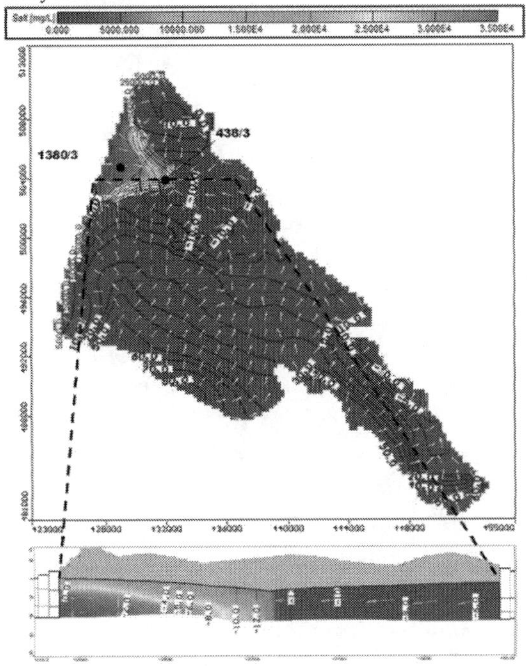

Fig. 4. Calculated piezometry and salinity for the moderate scenario, a) Plan view of salinity simulated in 2040 for Model layer 3, and b) Transversal section of simulated salinity.

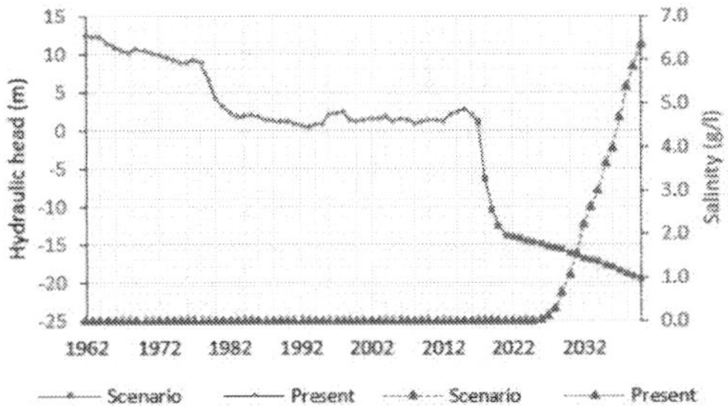

Fig. 5. Time series of Hydraulic head and salinity at piezometer 438/3 located near the ONEE well field.

4 Results: GALDIT method

The GALDIT method is an intrinsic vulnerability assessment developed by [11] to assess the vulnerability of coastal aquifers to seawater intrusion. It is a parametric system with rating scores and weights for six parameters (Table 1) combined to calculate the vulnerability index:

- G: Groundwater Occurrence;
- A: Aquifer Hydraulic Conductivity;
- H: Height of Groundwater Level below Sea Level;
- D: Distance from the Shore;
- I: Impact of the existing status of seawater intrusion in the area);
- T: Thickness of the aquifer.

Table 1. GALDIT Parameters and ratings

P*	W*	Ranking			
		Very Low	Low	Middle	High
		2.5	5	7.5	10
G	1	Bounded aquifer	Leaky confined	unconfined	confined
A	3	< 5	5 - 10	10-40	> 40
L	4	< 1	1 – 1.5	1.5 - 2	> 2
D	2	< 500	500-750	750-1000	> 1000
I	1	< 1	1 – 1.5	1.5 - 2	> 2
T	2	< 5	5 – 7.5	7.5-10	> 10

P*: Parameter
W*: Weight

The calculation of the GALDIT vulnerability Index is given by the sum of the six index maps at each cell (size: 250 m for our case study) following equation (1):

$$GALDI\text{-}index = \frac{\sum_{i=1}^{6} W_i R_i}{\sum_{i=1}^{6} W_i} \qquad (1)$$

Where R is the rating and W is the weight string, as indicated in table 1. The final GALDIT vulnerability index score varies from 5 to 10 and is divided into three vulnerability classes: high (>10), moderate (7.5 to 10), and low vulnerability (5 to 7.5).

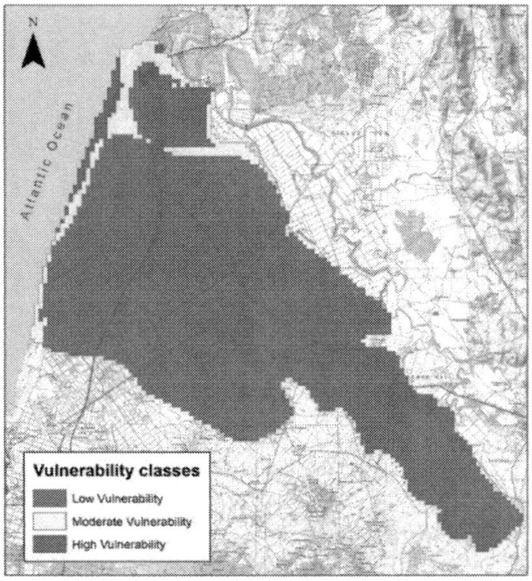

Fig. 6. GALDIT specific vulnerability map of groundwater (year 2014)

The results show that for 2014 (Fig. 6), the vulnerability classes obtained by this combination vary between 5 and 10. They are distributed in three classes corresponding to the degrees of vulnerability fluctuating from "Low" to "High". The dominant class is the "Low vulnerability", covering up to 93% of the study area. From Medium to High Vulnerability classes cover an area up to 5% to 2% respectively and are located in the north-western part of the study area. Moreover, these results were confirmed by the Numerical modeling of SWI. They reveal that the fringe littoral areas of the aquifer are the most affected by seawater intrusion, with a high risk in the north-western part of the study area.

5 Conclusion

Groundwater is the important source of freshwater in Rmel O. Ogbane coastal plain in the Low Loukkos basin (Morocco), where 98% of the domestic water and the whole industrial water depend on groundwater. Therefore, it is imperative to protect the groundwater aquifer from seawater intrusion in this area. To predict the extent of seawater intrusion in the future and provide helpful information for the protection of groundwater resources, a GALDIT vulnerability map and a 3D numerical model of density-dependent groundwater flow and miscible salt transport have been developed to assess the current extent of seawater intrusion in the study area. These two approaches are used to better understand coastal aquifer vulnerability to seawater intrusion for the Rmel-O. Ogbane aquifer in this semi-arid region of Morocco under CC and human activities.

The results show that the main effect of seawater intrusion in the Rmel-O. Ogbane aquifer will be the excessive over-pumping that will decrease the renewable resources, which with the pessimist scenario may be increased by 30% of present-day values. Also, the decline in recharge and the rise in sea level due to the CC accelerate saltwater intrusion into the aquifer system, which reduces the fresh groundwater resources. The simulation results show that the maximum extent (about 5.2 km) of seawater intrusion will increase in 2040 in the north-western sector. The water quality will be affected mainly in the ONEE pumping area, immediately adjacent to the seashore.

Moreover, these results were confirmed by the application of GALDIT method. It is capable of assessing the vulnerability to seawater intrusion of the coastal aquifer of Rmel. In 2014, the groundwater aquifer was characterized by a low vulnerability upstream and a high vulnerability downstream with an intense vulnerability of the marine intrusion in the coastal area and the Loukkos River's proximity. According to GALDIT method, the aquifer is highly vulnerable up to 1 km inland in the north-western part of the study area.

Indeed, this derived 2D and 3D maps can be used as a tool for a decision support system to assist better strategies for groundwater development, planning, and management of the coastal groundwater resources.

References

1. T. P. Barnett, D. W. Pierce, H. G. Hidalgo, C. Bonfils, B. D. Santer, T. Das, G. Bala, A. W. Wood, T. Nozawa, A. A. Mirin, D. R. Cayan, and M. D. Dettinger, Science (80-.). **319**, 1080 (2008)
2. IPCC (Intergovernmental Panel on Climate Change), (2007)
3. J. P. (Moroccan G. S. R. Thauvin, *Ressources En Eau (Tome 1) : Le Bassin Du Bas Loukkos* (Moroccan Geological Survey Report, Rabat, 1971)
4. M. J. El Hamidi, A. Larabi, M. Faouzi, and M. Souissi, Arab. J. Geosci. **11**, (2018)
5. M. Messaoud, *Hydrogéologie Du Bassin Du Bas-Loukkos (Larache-Ksar-El-Kbir), Province de Tétouan, Maroc.* (Montpellier, France, 1963)
6. M. Bonnet, *Méthodologie Des Modèles de Simulation En Hydrogéologie* (Nancy, France, 1982)
7. A. Arbai, *Etude Hydrogéologique et Modélisation de La Nappe de Rmel* (Casablanca, Morocco, 1999)
8. A. Larabi, *Note de Synthèse Hydrogéologique de l'aquifère de R'mel* (Rabat, Morocco, 2002)
9. M. J. El Hamidi, A. Larabi, M. Faouzi, and R. Essafi, Rev. Des Sci. l'Eau **32**, (2019)
10. G. Choubert and R. Ambroggi, *Note Préliminaire Sur La Présence de Deux Cycles Sédimentaires Dans Le Pliocène Marin Au Maroc* (1953)
11. A. G. Chachadi and J. P. Lobo-Ferreira, in *Proc. Work. Model. Hydrogeol.* (Proceedings of the Workshop on Modelling in Hydrogeology, Anna University, Chennai, 2001)

https://doi.org/10.1051/e3sconf/202129805003

Application of electrical Tomography for the hydrogeological study of the Fez-Taza corridor, Case of Bir Tam-Tam(Morocco)

Hassnae Faiz[1,*], *Anasse* Benslimane[1], *Mohamed* Chibout[1,3], *Mohamed* El Mokhtar[2,3], *Fatima Zahra* Faqihi[1], and *Youssef* Chahid[1]

[1]Laboratory Geosciences, Environment and Associated Resources, Faculty of Sciences Dhar El Mahraz, University Sidi Mohamed Ben Abdellah, Fez, Morocco.
[2]Laboratory of Geosciences, Faculty of Science, University Ibn Tofail, Kenitra, Morocco.
[3]Society AFRICA GEO-SERVICES, Fez, Morocco.

Abstract. The present work contributes to the multidisciplinary geological, hydrogeological, and geophysical exploration of groundwater in the Bir Tam-Tam region of the Fez-Taza corridor. Our main objective is to release water resources in our study area to supply drinking water in the region and irrigation of agricultural areas. Potential zones are located at the Lias fracturing zones. This study is interested in understanding the hydrogeological context of the study area through geophysical methods such as Electrical Tomography. The comparison of geological data, Drilling data, and the interpretation of the results of electrical Tomography have made it possible to highlight the geoelectric levels likely to constitute a potential aquifer and to locate possible structural accidents (faults) affecting the dolomitic limestone formations of Lias that could drain groundwater.

1 Introduction

The need for water in the world is increasing more and more with time; the resolution of this problem is mainly based on the development of science and techniques for groundwater research.

The present work is part of a better knowledge of groundwater resources in the Fez-Taza corridor. Non-detailed hydrogeological studies characterize this region; hence the objective is to improve hydrogeological knowledge and assess the potential of groundwater resources in this area which covers a geographical area where the water demand is constantly growing, especially for drinking water supply DWS for Taza, irrigation, and industrial activities.

2 General framework of the study area

2.1 Situation of the study area

- <u>From a regional point of view:</u> the study area is located inside the Fez-Taza corridor (Fig.1). The boundaries of this corridor are defined in the North by the outcrop limit of the essentially marly formations belonging to the pre-rifain layers; in the West, the limit of the basin of Fez-Meknes. From the WSW to the ENE, the boundary of the continuous outcrops of the liasic limestone of the Middle Atlas joins towards the NE that of the pre-rifain layers [4]

- <u>From a local point of view:</u> The surface of the study area is located in Bir Tam-Tam, at an estimated distance as the crow flies, about 40 km ESE of the city of Fez, 30 km NE of the city of Sefrou, 57 km SW of the city of Taza and 11 km S of the dam of Idriss Premier. Bir Tam-Tam is a commune of the province of Sefrou in the administrative region of Fez-Meknes in Morocco.

- <u>From an administrative point of view:</u> the study area is located in the Douar of Hmiddane in the rural commune of Bir Tam-Tam, province of Sefrou, region Fez-Meknes.

2.2 Climatic context

The Fez-Taza corridor is characterized by a rainfall regime that divides the year in two. The winter is rainy from November to April, and the summer is dry from June to September. Rainfall varies between 500 mm to the West and 700 mm to the east (Taza).

The average minimum temperatures vary between 11.1 °C (Fez, altitude 415 m) and 11°C (Taza, altitude 510 m); the average maximum temperatures are 24.6 °C in Fez and 24.1 °C in Taza.

2.3 Geological context

The Fez-Taza corridor constitutes the most eastern part of the South Rifain furrow.

The sedimentary series in the Fez-Taza corridor essentially includes:

- <u>Primary and Secondary:</u> The Primary consists of shales and sandstones, sometimes metamorphic represented in several buttonholes, the most important of which is Jbel Tazzeka.

* Corresponding author: hassnae.faiz@gmail.com

https://doi.org/10.1051/e3sconf/202129805003

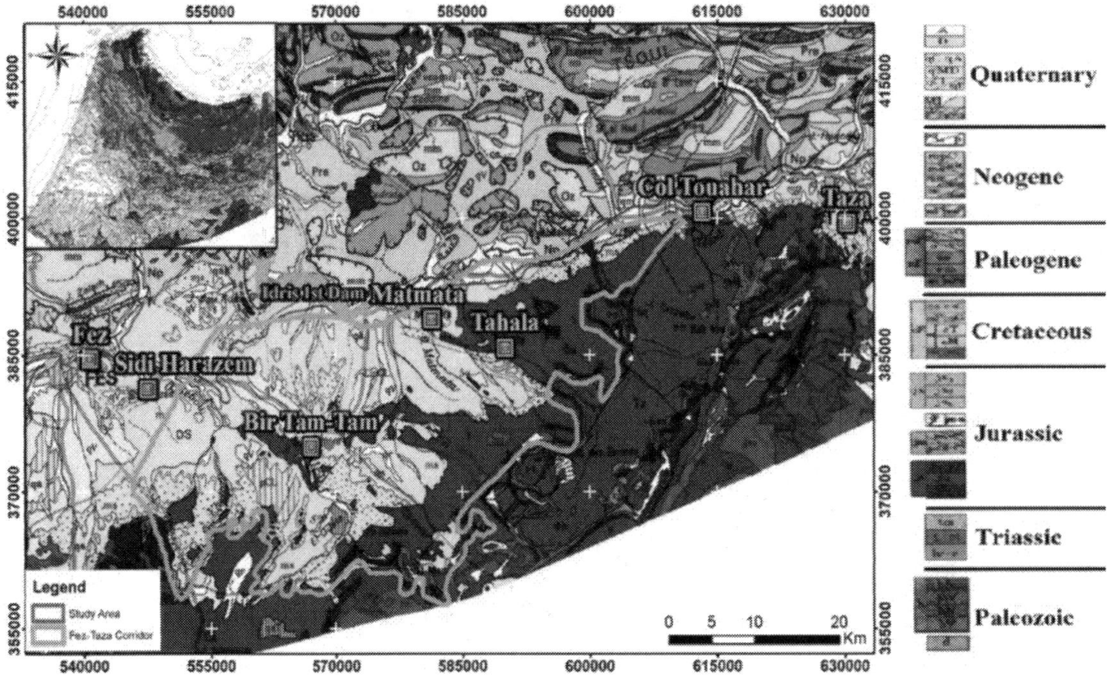

Fig. 1. a) Location map of the study area in the Fez-Meknes region, North of Morocco; b) Location map of the study area in the Fez-Taza corridor on the geological map of the Rif 1/500000

Then comes a permo-triassic series of continental and lagoonal red marls, discordant on the Primary, often accompanied by doleritic basalt flows [7].

The limestones and dolomites of the Lias come practically in concordance on the Permo-Trias. A massive series constitute the Lower Lias; then comes the middle Lias in dolomitic limestones and limestones with flint and limestone or sub-lithographic limestones.

- Miocene-Pliocene: The transgressive Miocene was deposited on a Causse that emerged since the Jurassic (Domerian), eroded, folded, and fractured. Two primary levels are distinguished: at the base, the detrital Helvetian (sandstone and marlstone), then the essentially marly Tortonian, with fine marlstone intercalations (called blue marl series).

1.4 Structural context

This region has never undergone significant tectonic movements after the Hercynian phases. The only deformations observed are faults due to epirogenic movements that occurred during the Second and Tertiary periods.

From a hydrogeological point of view, it is important to know the deep tectonics of the limestone extensions of the Middle Atlas Causse under the Miocene cover of the Fez-Taza corridor Lias limestones constitute the reservoir that contains the main groundwater resources in this sector.

2 Hydro-geophysical study by electrical Tomography

In order to follow the hydrogeological characterization of the study area in the horizontal and vertical directions of the water reservoir in the Bir Tam-Tam sector, a geophysical campaign by electrical Tomography was carried out in the region.

- Principle of electrical Tomography:

Electrical Tomography allows obtaining an "electrical image" of the subsurface, i.e., a pseudo section of the apparent resistivity versus depth. Therefore, it is highlighted to obtain a basement model where resistivity distribution varies vertically and horizontally along with the profile [3].

- Measurements:

09 electrical tomography profiles distributed in the surface of the study area. (ARES II geoelectrical Equipment).

- Data processing by Res2Dinv software:

The values obtained in the field are apparent resistivities. From these values, we try to find the calculated thicknesses and resistivities of the different bodies present [5,6].

The Res2Dinv software proceeds through the steps of the inversion method, which is displayed in (Fig. 3).

Fig. 2. Map of the location of Tomography profiles on the satellite image in the study area.

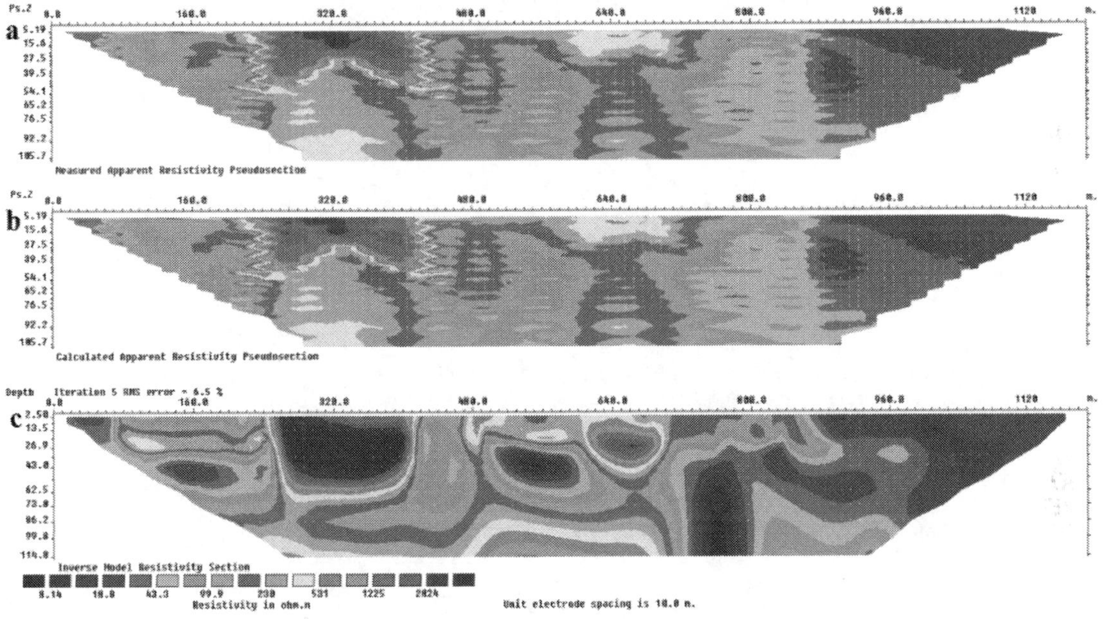

Fig. 3. Vertical sections of resistivities versus depth: a) Measured Apparent Resistivity Pseudosection; b) Calculated Apparent Resistivity Pseudosection; c) Inverse Model Resistivity Section corresponding to the calculated apparent resistivity Pseudosection.

3 Results and Discussions

The analysis of the first pseudo-sections in resistivity of the electrical tomography profiles BT1, BT2, BT3, BT4, and BT5, proves the presence of four anomaly zones according to the value of the true resistivity and presented in the following order:
- Resistive zones:
- Resistant zone A: corresponds to dolomitic limestone formations of the Lias (resistivity varies from 250 to 1500 ohm.m);
- Resistant zone D: corresponds to lacustrine limestone formations of the Pliocene (resistivity varies from 250 to 2000 ohm.m). For values higher than 2000 ohm.m, they correspond to the karstified dolomitic limestone formations (voids).
- Intermediate zone B: corresponds to formations made up of sandy marl at the top and sandstone-conglomerate (Burdigalian) at the base of Miocene age (resistivity varies from 25 to 200 ohm.m);
- Conductive zone C: corresponds to formations constituted by blue marls of Miocene age (resistivity varies from 3 to 20 ohm.m).

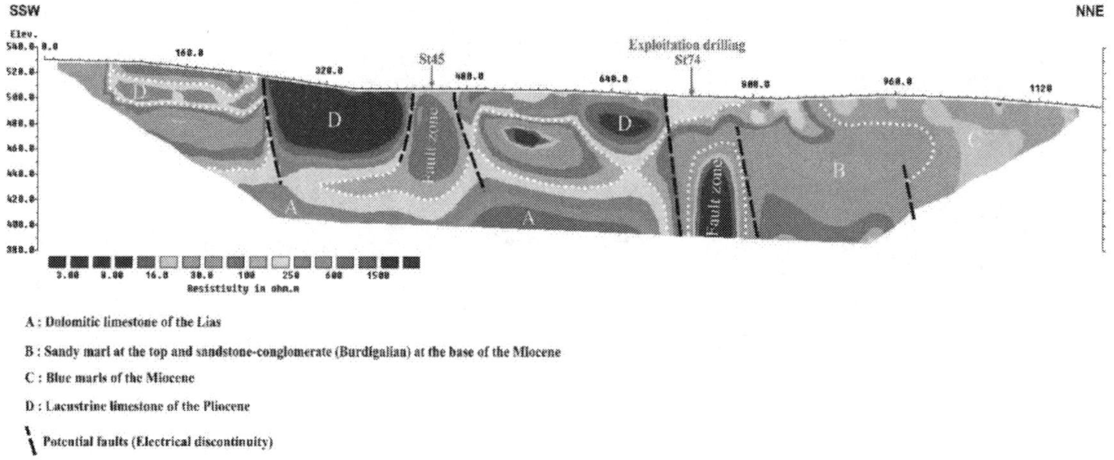

Fig. 4. 2D model of electrical resistivity distribution of the BT1 profile, with 10 m electrode spacing.

This pseudo-section shows the presence of 4 different formations mentioned in a vertical manner, which implies several electrical discontinuities of which two fault zones are well marked on the model (Fig. 4) (stations St45 and St74).

By analyzing the model, we also notice some electrical discontinuities that would probably correspond to faults that could represent targets for the location of future drillings for groundwater exploitation. Thus we were able to carry out the exploitation drilling (Pt=250m), which showed the presence of very fractured dolomite reaching a flow above 20l/s.

Conclusion

The geological maps, the lithological sections of some existing drillings, and the GIS used for the geological identification of the different geoelectric levels are highlighted by the electric tomography profiles for the hydrogeological characterization of the study area.

The analysis of the 2D models affirms the presence of a conductive superficial terrain attributed to the blue marls of the Miocene, which surmounts the intermediate geoelectric level appropriate to the sandy marls. The whole deposited on a resistant formation corresponding to the dolomitic limestones of the Lias on the tomographic profiles we notice the presence of local resistant superficial zones corresponding to the lacustrine limestones of the Pliocene (anomaly zone).

The electrical tomography method is a powerful method that has allowed us to study and recognize our land and execute by the end of the drilling of exploitation with essential flows. Hence, the recommendation is to encourage more geophysical techniques and develop the methodology of electrical prospecting.

References

1. J. Astier, *Geophysical applications in hydrogeology*, Masson edict. Paris 275P (1971)
2. M. Chibout, A. Benslimane, M. El Mokhtar, F. Faqihi, S. M. El Kanti, A. Ntarmouchant, 2019. Geophysical Contribution to the Determination of the Limit between Laayoune and Foum El Oued Aquifers, Origin of Supply Sources of Oued Sakia El Hamra, Laayoune Province, Morocco. International Journal of Geosciences, (2019). https://doi.org/10.4236/ijg.2019.1011058
3. T. Dahlin, 2D resistivity surveying for environmental and engineering applications. First Break 14:275–283. (1996) ; https://doi.org/10.3997/1365-2397.1996014
4. O. Fassi Fihri et al. Hydrogeological study of the "Fez-Taza corridor". Karst Hydrology (Proceedings of Workshop W2 held at Rabat, Morocco, April-May 1997). IAHS Publ. no. 247, 1998.
5. M.H. Loke, Barker RD Rapid least-squares inversion of apparent resistivity pseudosections by a quasi-Newton method. Geophysics Prospect 44:131–152. (1996)
6. M.H. Loke, Tutorial: 2D and 3D electrical imaging surveys. (1996-2002)
7. A. Michard, Elements of Moroccan geology, Notes and memoirs of the geological service of Morocco252, 408. (1976).

https://doi.org/10.1051/e3sconf/202129805004

Metallic trace elements contained in the sediments of Dayet Er-Roumi Lake (Morocco)

Sara El Ghizi [1,2,3*], *Sliman* Hssaissoune[2], *Mohammed* El Bouch[2], *Mohamed* Sadik[3], and *Mustapha* Hasnaoui[1]

[1]Environmental Engineering Team, Department of Biology. Faculty of Sciences and Techniques, Sultan Moulay Slimane University. BP. 523, 23 000 Beni Mellal, Morocco.
[2]National Laboratory for Pollution Studies and Monitoring (LNESP), Rabat-Morocco.
[3]National Center for Hydrobiology and Fish Farming, Azrou. HCEFLCD, Morocco.

Abstract. The contamination of aquatic ecosystems with metals remains a serious environmental problem of growing concern. Sediments are often studied as reservoirs or wells for many chemical pollutants. They are micropollutant traps; they also indicate the watercourse's historical pollution, lagoons, and lakes...Lake Dayet Er-Roumi (SIBE) is the only permanent natural lake in the Khemisset region (Morocco); this wetland is affected by several forms of pollution that are mainly linked to human activities (agricultural, domestic, industrial, etc.). The objective of our study is to contribute, through qualitative and quantitative analysis, to the assessment of lake sediment pollution by Spatio-temporal monitoring of certain toxic heavy metals such as Mn, Zn, Cr, Cd. These sediments are collected on the level of the lake's surface; the concentration of heavy metals was measured by microwave plasma atomic emission spectroscope (MP-AES). Metal analyses carried out at three measuring points during the winter season reveal a high concentration of Zn, Cr, and Mn at all stations during the winter season, which far exceeds the recommended standards, with a low level of Cd. The contamination factor "FC" and the degree of contamination "DC" reveal polymetallic contaminations dominated by two elements, zinc, and Chromium, which are the most worrying. These results clearly show that lake Dayet Er-Roumi is polluted; strategies to limit the lake's pollution must be implemented to avoid several ecological problems (fish mortality, etc.).

1 Introduction

Aquatic ecosystems cover 70% of the planet's surface. Pollution now threatens them because of human activities [1], particularly urbanization and industrialization... Among the many compounds emitted by human activities, heavy metals are one of the major pollution sources. These inorganic chemicals are being released into the environment as a result of human activities. Whether it is a specific spot of this environment (industrial and urban effluents, etc.) or a diffused one (runoff, dry and wet atmospheric deposition, etc.), These inorganic chemicals can penetrate aquatic ecosystems and integrate into the particle phase. These particles can settle on the bottom materials, where contaminants can accumulate over time [2]. Therefore Sediments can act in the long term as a reservoir of chemicals for the aquatic environment and organisms that inhabit the sediments or have direct contact with bottom materials. Since the sediment is a dynamic and essential compartment in aquatic ecosystems and plays an essential role for many animals and plants as a habitat or a nesting site, the exposure to certain substances contained in sediments could have a serious health risk to these organisms and also lead to critical or even dangerous situations affecting the ecological balance, and leading to the dysfunction of the trophic chain (low biodiversity, etc.) [3].

For many years, the metallic contamination of surface sediments (Mn, Ni, Cr, Cr, Cu, Zn, Pb, etc.) has attracted researchers' attention from very different horizons. It is one of the most threatening pollution aspects in aquatic environments. According to Forstner and Wittman (1981) [4], a significant fraction of heavy metals in the aquatic environment is reversibly associated with surface sediments, providing access to a more or less complete record of the state of contamination in aquatic environments. Therefore assessment of heavy metal concentrations in an aquatic system and their effect on organisms is essential.

As for other countries in the world, Morocco also does not escape the scourge of pollution. The Lake Dayet Er Roumi (Site of Biological and Ecological Interest) is the only permanent natural lake in the Khmisset region; it plays a significant socio-economic and ecological role, But this Daya has experienced fluctuations in its state, linked to climatic hazards on the onehand, and anthropic activities on the other.

Several studies [5] and [6] have shown that the water of Lake Dayet Er-Roumi is very polluted; they found strong mineralization (a high salinity, with a dominance of chlorides) and significant metallic pollution (Al, Mn,

* Corresponding author: sara.elg2014@gmail.com

and Pb ...) which reflects the direct influences of human activity (intensive and uncontrolled use of phytosanitary products, wastewater, domestic waste ...) on the ecological quality of the lake. A complete diagnosis of the current pollution situation is essential to safeguard this ecosystem. Our work's perspective consists of assessing the degree of pollution, evaluating the metal quality of the lake's sediment during the winterseason of the year 2019, and studying human activities' impact on the quality of the lake.

2 Material and Methods

2.1 Study area study

Dayet Er-Roumi Lake (33°45'N 06°12'W) is located 15 km southwest of Khémisset, situated in three rural communes: Ait Ouribel, Ait Ouahi, and AitHouderrane. It is characterized by a semi-arid climate with a maximum summer temperature of 38°C, a minimum winter temperature of 7°C, and a middlingrainy regime. It also has a 2 km long and a 400 m to 700 m width, with an area of about 90 ha and a maximum depth of 13.5 m in the center (according to the National Center for Information Exchange on Biodiversity of Morocco [7]), This lake is fed by groundwater and two tributaries (Fig. 1).

2.2 Samples

The choice of sampling sites has been made according to the juxtaposed activities; in total, there are five sampling stations (S1-S5), the three points in the lake (S1-S3) are representing the areas that are more likely to be affected by human activities (housing, hotel, etc.) and the other two points (the red sampling stations) are the stations where sampling is complex because there are plants and stones

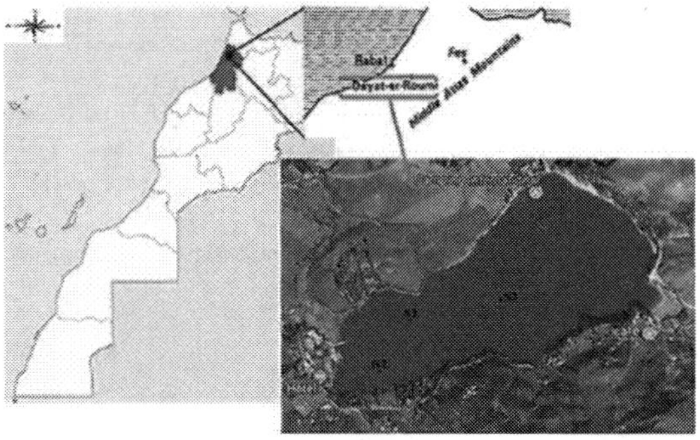

Fig.1. Study area and location of sampling stations.

2.3 Sample preparation and analysis

Sediment samples are collected on the surface layer of the lake using an Ekman bucket. The samples are transferred to clean plastic bags and then transported to a cooler (4°C). In the laboratory, these samples were dried in an oven for 24 hours at 80°C and sieved to obtain a powder with a diameter of ≤ 63 µm that has been stored in well-closed plastic vials to protect them against humidity. The vials were then stored in a cabinet in the dark and a cool place (20°C) until digestion. The method used for this digestion is total decomposition [8], 0.2g dry sediment samples were collected and placed in

Teflon tubes were previously washed with acid. For hot digestion, 5 ml of nitric acid, 2 ml of hydrogen peroxide, and 2 ml of hydrofluoric acid were placed in each tube and were put in a microwave oven at a temperature of 190°c for 45 min. after digestion of the samples, a 10 mL boric acid solution was added to the contents of thetubes, the final volume was reduced to 50 ml. The metals were measured by Microwave Plasma Atomic Emission Spectrometer (MP-AES). The validity of the analytical methods was verified by internal control and by an external control using an inter-calibration exercise [9].

2.4 Statistical Analysis

For a better results interpretation, we relied on statistical analysis, a calculation of Contamination factor "FC" and a Degree of contamination (DC). Metals are mainly associated with sediment particles (clays, carbonates, iron oxides and hydroxides, organic matter, etc.); it is, therefore, necessary to express the concentrations of metals according to a parameter related to the nature of the sediment.

2.4.1 Contamination factor "FC":

This normalization consists in expressing the ratio of the content of a given substance compared to that of the normalizing factor (We have used the crustal concentrations of Wedephol [10] and Pekey [11] as reference values (UCC-Upper Continental Crus)), allows defining the contamination factor "FC" of a given trace element in the sediment. This contamination factor is expressed by the following formula: $FC = Cx/Bg$

Cx: concentration measured for an element x, Bgx: Background for an element x

Contamination classes are defined for FC [12], [13] [14]: $FC < 1$ (absence to low contamination),

$1 \leq FC < 3$ (moderate contamination), $3 \leq FC < 6$ (significant contamination), $6 \leq FC$ (very high contamination).

2.4.2 Degree of contamination (DC):

Allows the estimation of a priori polymetallic contamination for each sampling point. It is the sum of the FC; it is calculated according to the following formula (Hakanson 1980): $DC = \sum FC$

This index is associated with 4 quality classes [12]: $DC < 6$ (low contamination),

$6 \leq DC < 12$ (moderate contamination), $12 \leq DC < 24$ (significant contamination), $24 \leq DC$ (very high contamination).

3 Results and discussion

3.1 Metallic contamination of sediments

The three Figures reveal the existence of metallic contamination characterized by significant Spatiotemporal variability. The highest average concentrations are those of Mn (1232.5 mg / kg) followed by Zn (993.94 mg / g) and Cr (102.424 mg / kg).

Concerning the heavy metal manganese (Fig.2), significant levels were recorded. The manganese concentrations vary between 1232.5 mg / kg and 684.96 mg/kg. The maximal concentration (1232.5 mg/kg) was observed at station 1, which indicates an anthropogenic activity (uncontrolled use of fertilizers and pesticides). The values recorded largely exceed the recommended standards [15] and [16]. These results show the vulnerability of the lake to extreme pollution.

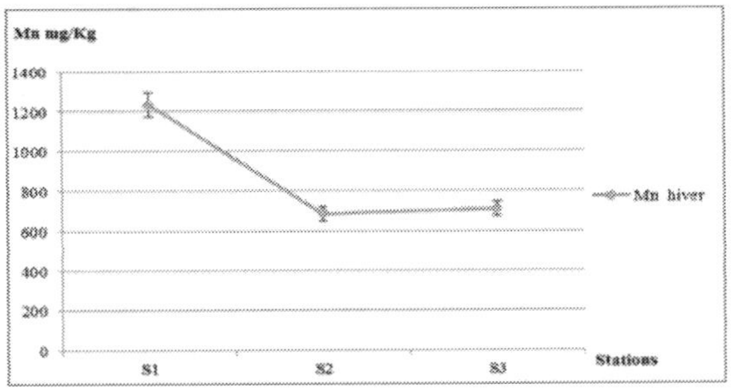

Fig.2. Spatio-temporal variation of Mn.

According to Figure 3, Zinc values reveal significant concentrations over the winter seasons; it varies between 993.94 mg/kg and 196.514 mg/kg; the maximal values were recorded during the winter at station 2. The high concentration of zinc is associated mainly with anthropogenic activities (agriculture, domestic waste, etc.). Our results show that the average zinc concentration far exceeds the recommended standards [15] and [16]. These high concentrations can have adverse effects on the aquatic environment.

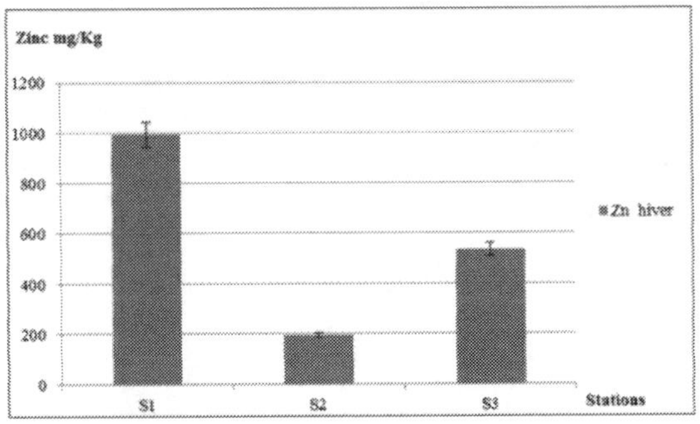

Fig.3. Spatio-temporal variation of Zn.

Chromium concentrations range from 102.75 mg/kg to 58.424 mg/kg (Fig.4). The monitoring at all stations reveals high concentrations during the winter season, particularly Station 3 is reaching a high concentration of 102.75 mg/kg and exceeds recommended standards [15] and [16]. This high concentration is likely due to the leaching of agricultural land showing the intense anthropogenic activity in the region and its impact on lake sediment quality.

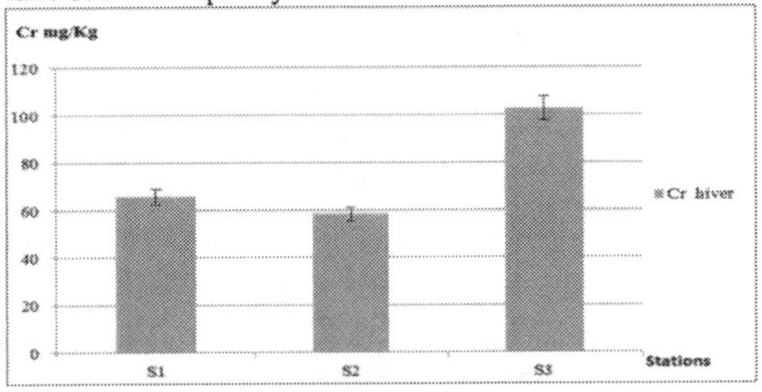

Fig.4. Spatio-temporal variation of Cr.

According to the results, we conclude that the sediments of lake Dayet Er-Roumi are polluted, the concentration of Mn, Zn, and Cr is very high in all measuring stations and vastly exceed the standards [15] and [16], this suggests that a significant part of these heavy metals may come from the inputs used, such as chemical fertilizers, pesticides (insecticides, fungicides, etc.), and may also come from domestic waste and landfills, etc. For Nickel, the concentrations at the stations studied are deficient and almost negligible. All concentrations recorded during the winter season at all stations are not detected (below the detection limit) by MP-AES.

3.2 Index and degree of sediment contamination

In order to have a quantitative idea of the sediments, additional analyses are carried out. The index approach also makes it possible to predict the extent of sediment pollution and identify priority polluted sites. It is found that the FC-Mn and FC-Cr are greater than 1 ($1 \leq$ FC-Mn < 3), which reflects an enrichment by these elements with moderate contamination. For the FC-Zn, the sediments of all sampling sites have FC greater than 6 ($6 \leq$ FC) except station 3, reflecting enrichment or even contamination by this element, which is considered extremely toxic to wildlife. The FC calculation shows that the lake's sediments are highly contaminated by the metals studied (Table 1).

For DC, the analysis of the results reveals polymetallic contamination dominated by two or more elements. The averages DC exceed the maximum class threshold defined by Hakanson (1980), indicating significant contamination for all stations with an average DC of 14.87; this value indicates significant polymetallic contamination (Table 1).

4 Conclusion

Overall, it appears that the levels of ETM show Manganese, Chrome, and zinc enrichment. This inventory reflects the direct influences of anthropogenic inputs. The various contamination indices calculated revealed situations of concern for several stations and several ETMs. Cases of polymetallic contamination dominated by two or more elements have been recorded (Zinc and Cr being the most worrying). All analyses show significant concentrations in the sediment. However, if the concentrations recorded do not give rise to immediate concern and cannot cause acute toxicity, it should be stressed that the eco-toxicological risk lies in the cumulative nature of heavy metals involved in bioaccumulation and bioamplification phenomena. Heavy metals can thus accumulate in low doses in various organs and reach the toxic threshold by altering either metabolic reactions in individuals or by causing demoecological disturbances in the populations of the affected ecosystems.

Table1. Contamination factor (FC) and degrees of contamination (DC) of the sites studied.

	Manganese mg/kg	**Zinc mg/kg**	**Chromemg/kg**	**Facteur of Contamination "FC"**			**Degree ofContamination "DC"**
				Fc Mn	**Fc Zn**	**Fc Cr**	
S1	1232.56	993.942	65.8225	2.338	19.112	1.880	23.33
S2	684.96	196.514	58.4245	1.299	3.7791	1.669	6.7471
S3	709.442	534.543	102.757	1.346	10.279	2.936	14.561
Average	875.656	574.999	75.668	1.661	11.057	2.162	14.87

References

1. C. Lévêque, *Aquatic Ecosystems. The Fundamentals*, Hachette (pp. 160). ISBN 2-01- 14-51264. (1996)
2. Canadian Council of Ministers of the Environment. 2001. *Canadian Sediment QualityGuidelines for the Protection of Aquatic Life* - Introduction, updated, in CanadianEnvironmental Quality Guidelines, 1999.
3. A. Lazhari, *Study of the contamination by metallic trace elements of the sediments of the Moulouya wadi and the Hassan II dam reservoir downstream of the abandoned Zeïda mine, Haute Moulouya*, P 70, (2013)
4. U. Förstner, G.T.W. Wittmann, *Metal pollution in the aquatic environment*. Springer-Verlag, Belin Heidelberg New York, 486 p . (1981)
5. S. El Ghizi, S. Hssaissoune, M. El Bouch, N. El Aadel, M. Sadik , M. Hasnaoui, *Assessment Of Metalic Pollution In Water Of Lake Dayet Er-Roumi (Morocco)*. In Proceedings of International conference Geo-IT and Water Resources 2020 (GEOIT4W'2020). Al- Hoceima, Morocco, (2020).
6. S. Ghizi, N. Aadel, M. Sadik, M. El Bouch, M.Hasnaoui, *The physicochemical characteristics and the pollution level of Dayet Er-Roumi Lakein Morocco*. E3S Web of Conferences 234, 00036 (2021).
7. M. MABUCHI, Datasheet on the SIBE (H9)Dayet Erroumi. National Center of Information exchange on the Biodiversity of Morocco, Convention on Biological Diversity, in,https://ma.chmcbd.net/manag_cons/esp_prot/ sibe_ma/sibe_cont_hum/dayet-Erroumi -h9-./si be_h9. (2006).
8. NM ISO 5667-15, Guidelines for the preservation and processing of sludge and sediment samples. (2012)
9. IAEA Marine environnement laboratory. MESL Report. IAEA/080/IAEA/MEL/56, Worldwide Regional intercomparison for determination of trace elements in polluted marine sediment. IAEA-356, Monaco.72 p.(2003)
10. WEDEPOHL K.H. *The composition of continental crus*", Goechimica and Cocmochimica Acta. Vol. 59, n°7 pp. 1217- 1232. (1995).
11. H. PEKEY, *The distribution and sources of heavy metals in Izmit Bay surface sediments affected by polluted straem*, Marine pollution bulletin. 52;1197-208. (2006)
12. L. Hakanson, *Ecological risk index for aquatic pollution control*. A sedimentological approach. Water Research, 14, 5, 975-1001 (1980).
13. U. Förstner, G.T.W. Wittmann. *Metal pollution in the aquatic environment*. Springer-Verlag, Belin Heidelberg New York, 486 p.(1981)
14. B. Rubio, M.A. Nombela, F. Vilas ,*Geochemistry of major and trace elements in sediments of the Ria de Vigo (NW Spain): An assessment of metal pollution*. Marine Pollution Bulletin, 40, 968-980. (2000).
15. Canadian Council of Ministers of the Environment. 1999. Canadian Sediment QualityGuidelines for the Protection of Life, in Canadian Environmental Quality Guidelines, 1999, Winnipeg, Council, Update 2001
16. OMEE (Ontario, USA) 1993. Guidelines for the protection and management of aquatic sediment quality in Ontario.

A comparative study to assess surface water's anthropic and natural vulnerability using the DKPR and RUSLE approaches. Case of the watershed of the Joumouaa dam (Coastal basin of Ghiss, North East Morocco)

Soukaina Mansour[1,*], *Taoufiq* Kouz[1], *Abdeldjalil* Belkendil[2], and *Hinde* Cherkaoui Dekkaki[1]

[1]Laboratory of Research and Development in Engineering Sciences. Research Team: Applied Geosciences and Geological Engineering. Department of Earth and Environmental Sciences. Faculty of Science and Technique of Al Hoceima-Abdelmalek Essaadi University. Morocco.
[2]Laboratory Water Resources Valorisation, Science and Technology Faculty, University of Tlemcen, Chetouane. Algeria.

Abstract. The salinization of surface water in a coastal context leads to a qualitative degradation of this resource by various sources of anthropogenic and natural pollution. In this context, we present the results of a comparative study using "DKPR" and "RUSLE" models to evaluate the degree of surface water vulnerability against pollution, especially in the sub-watershed of the Joumouaa dam, a hydraulic infrastructure providing drinking water for the Targuist city. The "DKPR" model adopted as a qualitative approach involves four parameters: Accessibility of the aquatic environment (D), Water functioning of the soil and subsoil (K), Physiography watershed (P), Rainfall erosivity (R). The final result is a resource vulnerability map obtained by combining index maps of these four parameters using remote sensing and GIS. The "RUSLE" model applied as a quantitative approach integrates five factors in a multiplying function, namely: rainfall erosivity (R), soil erodibility (K), slope length and steepness (LS), cover-management (C), and soil conservation practices (P) in a remote sensing and GIS environment. The analysis of the final vulnerability maps of the approaches mentioned above will be helpful support for water resource managers and decision-makers better identify areas of high risk and their protection.

1 Introduction

Surface water is considered the main water resource used for drinking water, irrigation, and industry. Thus, this resource exists in various forms available globally, namely springs, ocean ponds, and lakes [1]. Unfortunately, due to anthropogenic origin, natural and the effects of climate change will impact water use. This situation constitutes a significant problem whose degradation is expressed by salinization and contamination of waters that represents a constraint in coastal areas limiting the development of a region. In this context, the concept of "vulnerability" has been developed to determine which areas are more susceptible to pollution. Therefore, assessing surface water vulnerability is also necessary to understand the pressure on water resources [2] and design strategies that minimize these pressures.

In this study, the vulnerability to contamination of the waters of the sub-catchment of the Joumouaa dam, the source of the drinking water supply of the city of Targuist is assessed using the DKPR approach developed by the bureau of geological and mining research (BRGM) [3]. This approach has been widely used by many researchers [4, 5]. Thus, they obtained relevant results. In addition, the other approach, the Revised Universal Soil Loss Equation (RUSLE) [6], adjusted to local conditions. Its use is widespread in the Moroccan Rif to provide estimates of soil loss [7, 8].

The main objective of this paper is to apply a comparison between two approaches, one qualitative "DKPR"; to assess the vulnerability to water contamination of the Joumouaa dam of anthropogenic origin, and the other quantitative "RUSLE"; aims to map areas subject to a risk of water erosion of natural origin, to analyze the control factors and quantify the loss of soil at regional scale in a GIS and remote sensing environment. The results obtained are auspicious for optimal management of the quality of water resources and the protection of drinking water catchments.

2 Materiel and methods

2.1 Study Area

The studied basin corresponds to the sub-catchment of the Joumouaa dam, the main hydraulic infrastructure providing drinking water to the city of Targuist. It is part of the upstream zone of the Ghiss river basin. It is located

* Corresponding author: smansour@gmail.com

https://doi.org/10.1051/e3sconf/202129805005

in the northeast of the central Moroccan Rif, a few kilometers from the city of Al-Hoceima (Fig. 1.). It covers a total area of 52.07 Km^2 with a perimeter of 34.5 km. From a topographic point of view, the study area is characterized by an essential altitudinal variation going from 931 m at its outlet to 2051 m on the highest summit.

The prevailing climate is semi-arid to humid with a hot, dry summer and a wet, rainy winter. Rainfall is characterized by seasonal and annual irregularities with an average annual rainfall of 340 mm [9]. On the other hand, the highest average monthly temperature is 28°C, while the lowest is 7°C [10].

The region's hydrogeology characterized the coastal aquifer of Ghiss-Nekkour, which circulates below the studied sub-basin. This aquifer plays an important role in the agriculture and drinking water supply of the Al Hoceima city and its hinterland [11]. The bound waters are subject to challenges that threaten their sustainability due to salinization related to marine intrusion and anthropogenic pollution, and overexploitation resulting from the increasing demand for water due to the socio-economic development that the region as a whole is experiencing [11-13].

From a hydrological point of view, the study area is drained by Oued Joumouaa, and fed by the tributary of river Hmayed.

Geologically, the sub-catchment area is located in the rif a domain where three structural domains outcrop: The flysch domain is represented by the Jbel Tizirene nappe, composed of marly-calcareous-sand stone flysch, and by the Jbel Chouamat nappe represented by schistose flysch. The formations of the external zones cover almost all the studied territory; they are represented by the unit of Ketama and the unit of Beni Bou Yacoub. The internal area is mainly constituted by the unit of the limestone ridges with massive non-metamorphic limestones of the Jurassic age [14]. Moreover, the quaternary formations appear only in the downstream part of the studied sub-basin.

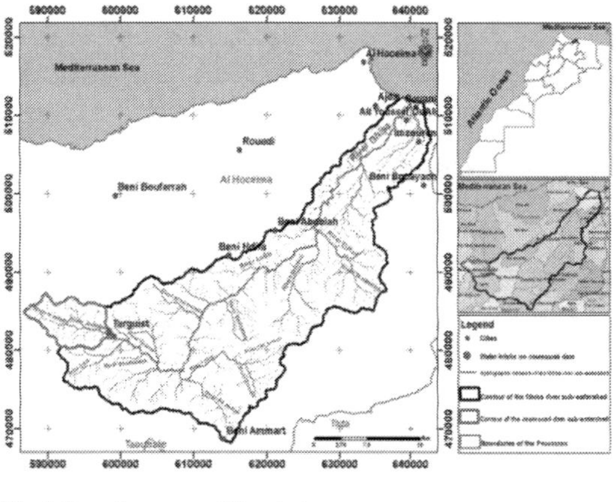

Fig.1. Location map of the study area

2.2 Principle and application of the DKPR and RUSLE approach

2.2.1 Principles of DKPR Model

The DKPR approach is based on the principle of parametric methods, using a system of weighted classes. It consists of evaluating the vulnerability of surface waters to contamination. The criteria used in the method are:

Aquatic Accessibility (D): Represents the hydraulic distance from each point to the nearest stream in its topographic basin. The class value assigned to this factor is inversely proportional to the distance "D" Table 1.

The hydrological functioning of the soil and subsoil (K): This parameter provides information on the partition between infiltration water and runoff water. Thus, the determination of this factor is done through the evaluation of the classes of three parameters, namely: the index of development and persistence of the networks "IDPR", the index of battance "IB" and the occupation of the ground "Os" Table 1. According to the equation below proposed by the DKPR method:

$$K = a\ IDPR + b\ IB + c\ \ Os \qquad (1)$$

With:
K: Index referring to the water functioning of the soil and the subsoil;
IDPR: Index class for network development and persistence index;
IB: Slaking index class;
Os: Land use class.
With: a = 0.4; b = 0.2 and c = 0.4 for a+b+c = 1

The physiography of the watershed "P": This parameter governs the runoff of surface water through the intensity of the slopes "Pi" and their shape (in relation to the curves) "Pc" Table 1, knowing that:

$$P = d\ Pi + e\ Pc \qquad (2)$$

With: d = 0.8 and e = 0.2 for d + e = 1
P: Watershed physiography criterion;
Pi: Slope intensity class;
Pc: Slope curvature class.

Rainfall erosivity "R": This factor is related to the intensity of rainfall. Thus, Renard and Freimund [15], propose a simplified formula integrating only the average annual rainfall P (in mm) Table 1:

$$R = 0.04830\ P^{1.610} \qquad (3)$$

Where:
R: Erosivity parameter;
P: Annual precipitation (mm).

The final degree of vulnerability "Vr" is the result of the sum of the products of each class by its weighting factor according to the following expression:

$$Vr = f\,D + g\,K + h\,P + i\,R \qquad (4)$$

With:
Vr: Vulnerability index of the resource;

https://doi.org/10.1051/e3sconf/202129805005

D: Class related to the accessibility to watercourses;
K: Class related to the hydric functioning of the soil and subsoil;
P: Class related to the slopes;
R: Rainfall erosivity class;
Where f, g, h and i are weighting factors: f = 0.3; g = 0.4; h = 0.2 and i = 0.1.

Table 1. Matrix generating the partial parametric indices [3]

Parameter	Degree of vulnerability and classes				
	Very High	High	Moderate	Low	Very Low
	4	3	2	1	0
D	0 - 50 m	50 - 100 m	100 - 200 m	200 - 500 m	>500 m
K	3.2 - 4	2.4 - 3.19	1.6 - 2.39	0.8 - 1.59	0 - 0.79
P	3.2 - 4	2.4 - 3.19	1.6 - 2.39	0.8 - 1.59	0 - 0.79
R	> 5000 mm	4000 to 5000 mm	3000 to 4000 mm	2000 to 3000 mm	0 to 2000 mm

Then, each pixel in the resource vulnerability map is assigned a value that determines its degree of resource is associated with a value determining its degree. Table 2 shows the classification of these vulnerability values.

Table 2. Classes of resource vulnerability indices [3]

Vulnerability class	Value of the Index Vr	Degree of vulnerability
4	3.2 – 4	Very High
3	2.4 – 3.19	High
2	1.6 – 2.39	Moderate
1	0.8 – 1.59	Low
0	0 – 0.79	Very Low

2.2.2 Principles of RUSLE Model

The revised universal soil loss equation [6] is an improved version of the USLE model of Wischmeier and Smith [16]. This empirical model combines the five significant factors controlling water erosion. It has been recognized as the most commonly used model for quantifying soil losses due to water erosion and spatially locating high priority areas for intervention in the study area.

The Universal Soil Loss Equation (USLE) is written:

$$A = R.K.LS.C.P \qquad (5)$$

Where:
A: Annual soil loss in t/ha/year;
R: Climatic aggressiveness factor in MJ.mm/Ha.H.year;
K : Soil erodibility expressed in t.ha.H/ha.MJ.mm;
LS: Factors of slope inclination and length (dimensionless);
C : Vegetation cover factor (dimensionless);
P: Factor of anti-erosion practices (dimensionless)

Rainfall erosivity (R): This factor presents the impact of the erosive force of rainfall on the soil. According to [16], the estimation of this index requires the determination of the kinetic energies and the maximum intensity over thirty consecutive minutes of the raindrops of each shower. Due to the lack of this type of data, some authors have developed alternative formulas that do not require detailed rainfall records but consider the average monthly and annual intensities to determine the R factor [17, 18]. For this, we choose the formulas of Rango and Arnoldus [18] it is given below:

$$\log R = 1.74 \log \sum_{i=1}^{12} \left(\frac{P_i^2}{P}\right) + 1.29 \qquad (6)$$

With:
Pi: Average monthly precipitation;
P: Average annual precipitation;
R: Climatic aggressiveness in MJ mm/ha H year.

Soil erodibility (K): expresses its sensitivity to water erosion and depends on its intrinsic properties such as its texture, structure, and permeability. Faced with the lack of these data types, we have developed the erodibility map based on previous work according to soil types characterizing the sub-basin studied [19, 20].

Slope length and steepness (LS): influences the rate of sediment loss, which is related to a significant increase in flux resulting from an increase in slope length and slope steepness [21]. Thus, the LS factor was calculated using the formula of Wischmeier and Smith [16], with modifications by [22] (Eq. 10).

$$LS = \left(\frac{L}{22.13}\right)^m + (0.065 + 0.045S + 0.0065S^2) \qquad (7)$$

Where S is the slope gradient in %; L is the length of the slope in meters. L = flow accumulation × DEM spatial resolution. Otherwise, the value of "m" varies between 0.2 and 0.5 depending on slope percentage. A value of 0.2 is attributed to areas with slope < 1%, values of 0.3 and 0.4 correspond, respectively, to areas with slope percentages that vary between 1–3 and 3–5, and a value of 0.5 is attributed to areas with slopes > 5% [16-22].

Vegetation cover factor (C): The index (C) represents the ratio of soil loss for a plot with a given vegetation cover and cropping practices to the rate of soil loss for a plowed and left bare plot. The generation of the raster layer of the C factor from the NDVI is written as follows [23-25]:

$$C = exp\left[\frac{-\alpha\ NDVI}{-\beta\ NDVI}\right] \qquad (8)$$

Where:
C: Vegetation cover factor; NDVI = (R5 - R4) / (R5 + R4); α and β are constants (α = 2 and β = 1) [26].

Soil conservation practices (P): Contour cropping, bench planting are very effective water and soil conservation practices. According to [16], the values of (P) are less than or equal to 1. The value of 1 is assigned to land on which there are no erosion control practices.

2.3 used Database

Several types of data from different sources were used:

- Digital Terrain Model (DTM) of 30 m resolution;
- Landsat 8 Oli images with a resolution of 30 m;
- Geological map of the rif chain at 1/500 000 scales provided by the ministry of energy and mines and sustainable development (geology department) Morocco;
- Pedological map of the rif and the oriental brought to 1/500 000 and provided by the ministry of agriculture of rural development and water and forests;
- Rainfall data from stations located in and around the study area for an observation period between 1989 and 2018 were obtained from the Worldclim database in raster format (ESRI grids) with a resolution of 2.5 minutes (about 21 km^2).

3 Results and discussions

3.1 DKPR approach

At the end of the application of the DKPR approach to evaluate the vulnerability of the waters of the sub-catchment of the Joumouaa dam to pollution, the final map of the intrinsic vulnerability of the waters of the studied sub-basin is represented as a qualitative map (Fig. 2.). It shows that the dominant class being of a moderate degree. It covers up to 61.22% of the total area of the studied territory. The rest of the studied area is represented by very low, low, and high classes occupying 25.31%, 13.38%, and 0.09 %. Thus, the confrontation of the four index maps obtained showed that the final map resulting takes the general appearance of the two index maps related to the criteria D and P. In this sense, the parameter "D" is considered one of the most important because it contributes most to the vulnerability of surface water. In addition, the most stable parameter is R.

Comparisons can be made with previous work undertaken in the Moroccan rif watersheds using the same approach [4, 5]. The results obtained are similar to those found in these recent studies.

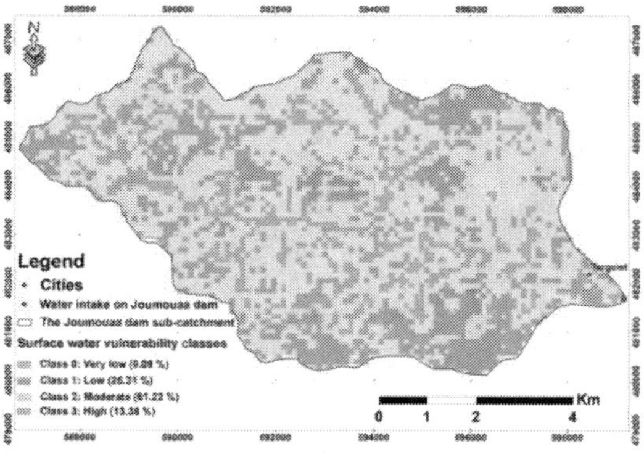

Fig. 2. Vulnerability of the surface waters of the Joumouaa dam sub-catchment

3.2 RUSLE approach

Overlaying the raster data layers of the model factors (R, K, LS, C, and P) were organized under the step of a grid with a cell size of 30 m × 30 m in a GIS environment provides a quantitative map estimating soil loss in the studied basin per spatial unit. These factors vary as follows:

- Climate Aggressiveness (R) Map: from 76.92 to 78.73 MJ mm/ha H year.
- Topographic factor map (LS): from 0 à 42.79.
- Soil erodibility map (K): from 0.23 to 0.46 t ha H/ha MJ mm.
- Vegetation cover map (C): from 0.07 to 1.
- Soil conservation practices (P): no practice is carried out at the level of the study area, and for that, we considered this factor is equal to 1.

The final result reveals an average annual soil loss ranging from 0 to 807.14 t/ha/year with an average of 16.71 t/ha/year (Fig. 3.). In addition, the soil loss map was grouped according to a classification adopted by FAO [27]. The areas with soil loss between 0 and 5 t/ha/year and represent 84% of the basin area and are distributed throughout the basin and generally follow the spatial distribution of low altitude and low slope areas, which highlights the effect of the topographic factor, especially the slope, in the water erosion process.

The areas with erosion rates ranging from 5 to 50 t/ha/year occupy 6.69%. The areas with a high risk of erosion, containing the classes with erosion rates between 50 and 100 t/ha/year, represent 4.48% of the studied territory. The areas with a very high erosion risk with values >100 t/ha/year also represent 4.74% of the area and follow the very steep slopes of the watershed.

The analysis of the results obtained with the RUSLE model highlights a medium vulnerability to water erosion, reflecting different erosion factors, mainly the topographic factor LS.

A comparison of the results obtained with previous works is necessary. Thus, the value of average annual soil loss (16.71t/ha/year) obtained in the studied basin is close to the values obtained in other similar environments [28, 29] and lower than those obtained by others in the same geographical context [7, 8].

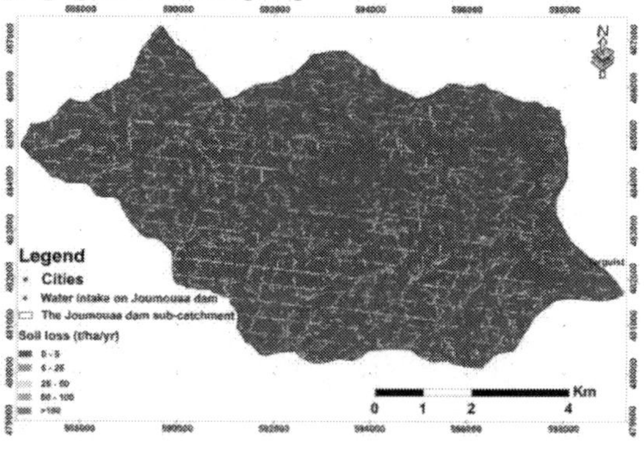

Fig. 3. Spatial distribution of soil losses

https://doi.org/10.1051/e3sconf/202129805005

Conclusion

The evaluation of the vulnerability of anthropic and natural origin at the level of the Joumouaa sub-catchment using the RUSLE and DKPR models in a GIS and remote sensing environment has highlighted the areas most exposed to erosion, thus polluting the responsible factors in order to ensure qualitative and quantitative management of the resource.

In this study, the results obtained from the DKPR model show that the dominant class being of a moderate degree of vulnerability where the parameter "D" contributes the most to the vulnerability of surface water. Concerning the RUSLE approach, the average annual soil erosion rate in the studied watershed was estimated at 16.71 t/ha/year, mainly due to the influence of the topography. Thus, both approaches show a medium degree of vulnerability. In addition, these results could be used as an aid to decision-makers and planners when making decisions in water and soil management and preservation.

The application of the DKPR and RUSLE approaches, despite some limitations, is beneficial for mapping areas at high risk of pollution and erosion, even in the absence of data. Both methods could be improved by obtaining more accurate data for the DKPR and RUSLE factors, such as precipitation data to better estimate the erosivity of precipitation and soil data to determine the K factor.

References

1. R. Bhateria, D. Jain.Water quality assessment of lake : a review. Sustain Water Resour Manag **2**, 2 (2016)
2. A. Anandhi, N. Kannan. Vulnerability assessment of water resources-translating a theoretical concept to an operational framework using systems thinking approach in a changing climate: case study in Ogallala Aquifer. J Hydrol **557**, 460–474 (2018)
3. D. Douay, E. Lardieg. Délimitation des aires d'alimentation des captages prioritaires du bassin Adour Garonne : méthodologie de cartographie de la vulnérabilité intrinsèque des captages d'eau superficielle. Agence de l'Eau ADOUR - GARRONE, ANTEA -GINGER Environnement & Infrastructures – CALLIGEE, 28 (2010)
4. I. Hilal, J.S. El Messari, J.G. Moreno Navarro, M.R.A. Mansour, H. M. Zerrouk. Application du modèle DKPR pour la cartographie de la vulnérabilité des eaux de surface à la contamination : cas du barrage 9 avril 1947 (Maroc septentrional). Revue des sciences de l'eau **31**, 1 (2018)
5. S. Mansour ,T. Kouz , M. Thaiki, A. Ouhadi , H. Mesmoudi, M. H Zerrouk, T. Mourabit, H. Dakak & H.Cherkaoui Dekkaki. Spatial assessment of the vulnerability of water resources against anthropogenic pollution using the DKPR model: a case of Ghiss-Nekkour basin, Morocco. Arab.J.Geosci **14**, 8 (2021)
6. K.G. Renard, G.R. Foster, G.A. Weesies, D.K McCool, D.C Yoder. *Predicting soil Erosion by water: a guide to conservation planning with the Revised Soil Loss Equation(RUSLE)* – Agriculture Handbook. USDA, Washington, 703 (1997)
7. A. Sadiki, S. Bouhlassa, J. Auajjar, A. Faleh, J.J. Macaire. Utilisation d'un SIG pour l'évaluation et la cartographie des risques d'érosion par l'Equation universelle des pertes en sol dans le Rif oriental (Maroc): cas du bassin versant de l'oued Boussouab. Bull l'Inst Sci Rabat , **26** (2004)
8. L. Khali Issa, K. Ben Hamman Lech-Hab, A. Raissouni, A. El Arrim. Quantitative Mapping of Soil Erosion RiskUsing GIS/USLE Approachat the KalayaWatershed (North Western Morocco). J. Mater. Environ. Sci **7**, 8 (2016)
9. Niazi S, Snoussi M, Foutlane A. Impact des aléas climatiques sur la qualité des eaux d'un hydrosystème aménagé en zone semi-aride : cas du bassin-versant du Nekkor (Maroc). Sécheresse (Montrouge) **16**, 3 (2005)
10. N. Arrebei, M. Sabir, M. Naimi, M. Chikhaoui, D. Raclot. Reconstitution des données historiques et diagnostic de l'état actuel des aménagements antiérosifs dans le bassin versant Nekor. Rev. Mar. Sci. Agron. Vét **7**, 2 (2019)
11. T. Kouz, H. Cherkaoui Dekkaki, S. Mansour, M. Hassani Zerrouk, T. Mourabit. Application of GALDIT Index to Assess the Intrinsic Vulnerability of Coastal Aquifer to Seawater Intrusion Case of the Ghiss-Nekor Aquifer (North East of Morocco). In: Calvache M, Duque C, Pulido-Velazquez D (eds) Groundwater and Global Change in the Western Mediterranean Area. Environmental Earth Sciences. Springer, Cham, 169–177(2017)
12. T. Kouz, S. Mansour, T. Mourabit, H. Cherkaoui Dekkaki. Application of the "PRK Plus" approach to evaluate the sensitivity of the Coastal Aquifer of Ghiss-Nekkour against pollution (North East of Morocco). La Houille Blanche, 5–6 (2019)
13. T. Kouz, S. Mansour, H. Mesmoudi, H. Dakkak, H. Cherkaoui Dekkaki. Assessment of groundwater vulnerability to pollution as part of integrated management in coastal area. Case of Ghiss-Nekkour basin (North East of Morocco). La Houille Blanche **106**, 2 (2020)
14. A. Michard, A. Chalouan , H. Feinberg, B. Goffé, Montigny R. How does the Alpine belt end between Spain and Morocco. Bull. Soc. Géol **173**,1 (2002)
15. K.G. Renard, J.R. Freimund. Using monthly precipitation data to estimate the R-factor in the revised USLE. Journal of Hydrology **157**,1–4 (1994)
16. W.H. Wischmeier, DD. Smith. *Predicting Rainfall Erosion Losses - A Guide to Conservation Planning.* Agriculture Handbook. Department of Agriculture Science and Education Administration, Washington, District of Columbia USA. US **537**, 58 (1978)
17. H.M.J Arnoldus. Methodology used to determine the maximum potential average annual soil loss due to sheet and rill erosion in Morocco. Soils Bulletins FAO **34**, 39-48 (1977)

18. A. Rango, H.M.J. Arnoldus. Aménagement des bassins versants. In : Cahiers techniques de la FAO 1–11(1987)
19. A. El Garouani, H. Chen, L. Lewis, A. Tribak, M. Abharour. Cartographie de l'utilisation du sol et de l'érosion nette à partir d'images satellitaires et du Sig Idrisi au Nord-Est du Maroc, Teledetection, Editions des Archives Contemporaines / Editions scientifiques GB / Gordon and Breach Scientific Publishers **8**, 3 (2009)
20. H. Mesrar. Modélisation, quantification et définition des facteurs qui contrôlent le risque de l'érosion hydrique. Cas du bassin versant de l'oued Sahla, Rif central, Maroc. Thèse de doctorat en Géologie, 369 (2016)
21. BP. Ganasri, H. Ramesh. Assessment of soil erosion by RUSLE model using remote sensing and GIS - a case study of Nethravathi Basin. Geosci Front **7**,953–961(2016)
22. A. Bizuwerk, G. Taddese, Y. Getahun. Application of GIS for modeling soil loss rate in Awash Basin, Ethiopia. International Live stock Research Institute, Addis Ababa, Ethiopia (2008)
23. P. Zhou, O. Luukkanen, T. Tokola, J.J. Nieminen. Effect of vegetation cover on soil erosion in a mountainous watershed. Catena **75**, 3 (2008)
24. M. Kouli, P. Soupios, F.Vallianatos. Soil erosion prediction using the revised universal soil loss equation (RUSLE) in a GIS framework, Chania, North western Crete, Greece. Environ Geo l**57**, 483–497 (2009)
25. V. J. Markose, K.S Jayappa. Soilloss estimation and prioritization of sub-watersheds of Kali River basin, Karnataka, India, using RUSLE and GIS. Environ Monit Assess **188**, 225 (2016)
26. J.M. Van der Knijff, R.J.A. Jones, L. Montanarella .Soil erosion risk assessment in Europe, EuropeanSoil Bureau Research Report EUR 19044 ENp, 34 (2000)
27. FAO P UNESCO. Metodología provisional para la evaluación de la degradación de los suelos. Roma (1980)
28. A. Sadiki, A. Faleh, J.L. Zezerze, H. Mastass. Quantification de l'erosion en nappe dans le bassin versant de l'Oued Sahla, Rifoccidental Maroc. Cahiers Géographiques, **6** (2009)
29. A. Zouagui, M. Benmansour, N. Amenzou, A. Nouira, M. Sabir, H. Benjelloun, H. Marah, A. Benkadad. Application la technique de 137Cs à l'estimation de l'érosion hydrique dans le bassin versant de Moulay Bouchta, Rif occidental, Maroc. . Rev. Mar. Sci. Agron. Vét, 1 (2012)

AUTHOR INDEX

EDP Sciences – Web of Conferences
17, Avenue du Hoggar
Parc d'Activité de Courtabouf
BP 112
F-91944 Les Ulis Cedex A
France

ISBN 978-1-7138-3487-8